PUNK ROCK AND UFOs

STRANGER THAN FICTION

MIKE DAMANTE

First Printing, 2020

ISBN 13: 978-1-7344198-4-9

Beyond The Fray Publishing, a division of Beyond The Fray, LLC, San Diego, CA

BEYOND THE FRAY

Publishing

www.beyondthefraypublishing.com

par·a·nor·mal– *denoting events or phenomena that are beyond the scope of normal scientific understanding.*

DEDICATIONS

This book is dedicated to those who are willing to let their imaginations and curiosity run wild.

"Yes, the Metahuman Thesis! More likely than not, these exceptional beings live among us, the basis of our myths. Gods among men upon our little blue planet here!" – Lex Luthor, *Batman v Superman: Dawn of Justice*

There was once a simpler time, a time void of the everyday absolutes that humanity takes for granted today.

An ancient time when the structures of society and the infrastructures of habit were being built and learned as they went. The founding fathers of literally everything were slowly shaping the groundwork for what we would come to know as commerce, agriculture, beliefs, societal roles, and life as we would experience it today. There were civilizations all over the world that were all learning to survive, and eventually the select few would thrive.

Before the idea of governments and kingdoms was formed, there was religion used to keep people in check, and prior to the establishment of religions, there was simply life as they knew it. "Religion" at that time was a way of life that would later be passed down and altered to form dogmatic belief systems and rules.

The ancient Sumerian texts, which were the first recorded accounts in the history of humanity, have been dissected and studied for years, with various theories stemming from them, but what is not indisputable is that these tablets that were used to write their texts were a valuable commodity. Why would the Sumerian people waste these tablets to write simple fables and fiction rather use them to record their history for the rest of the world to discover? How life was then, how it was viewed, and how it was recorded remains one of the great mysteries, but those early texts were a template for future generations to build off, as there are connections from those texts to Egyptian mythology and early Christian stories.

The core of theology, mythology, and future fables, science fiction, pop culture, comic heroes and common folklore all stem from these very early texts. What we would consider to be fiction or impossible today possibly could have existed, but through a different lens back then.

Could the idea of Gods exist? Or the idea that Gods walk among mortal men and have since the dawn of time? Is the idea of a Superman really that far-fetched?

Through researching the topics of UFOs, cryptozoology and the realm of the unexplained, the phrase "stranger than fiction" has always come to mind when trying to describe some of these cases to the everyday people not initiated into this weird paranormal club.

The previous Punk rock and UFOs book *True Believers* looked at the scope of the unexplained through those who study it, and how it connects to popular culture and our belief systems, and also was a call to arms to the common folk to take a newfound interest in these topics to help better understand each other and advance humanity. *Stranger Than Fiction* has a common goal: how do we take these tales of UFOs, alternate universes, mythical monsters, sky beings, cryptids, and other themes that sound like they were taken straight out of a comic book, and present a rational possibility that there could be some truth to these stories?

There are books that do a better job than I can in terms of detailing ancient astronaut theories, dissecting mythology and civilizations, and dissecting folklore. Jacques Vallee's *Passport To Magonia* is a brilliant study of cases that ties the UFO phenomena to folklore, and Jeffrey Kripal's *Mutants and Mystics* digs deep into the paranormal origins that the creators of our science fiction and pop culture legends came from. So what we have here is an idea to try to normalize the paranormal.

It is a hard sell to most to rebel against their previously held core beliefs and values, which was one of the early ideas of *Punk rock and UFOs: Cryptozoology meets Anarchy*. For starters, I was once of the firm belief that science as we know it was the be-all, end-all of every argument, and all religions were mere fairy tales twisted to keep down the uneducated and opiate the masses. Now, I've evolved to the idea that there is something there in religious texts (what that is exactly is still to be determined), and our traditional science is no longer the argument ender, as there are so many things our science simply cannot explain, control or solve, so it is acceptable to open our minds to other possibilities. The nuts-and-bolts approach to the study of UFOs is slowly being replaced by a social shift into the belief of consciousness, which I will jump right into in the first chapter. The quotes in the book were carefully selected from interviews I've done over the past few years for my website punkrockandUFOs.com, exclusive interviews for this project, and research for this book.

Ultimately, these books could be considered persuasive texts in the sense that they are trying to challenge a way of thinking. I appeared on *Coast To Coast* with

George Knapp, which was a huge honor to be interviewed by a respected investigative journalist on arguably the biggest platform for the paranormal. Towards the end of the interview, Knapp made the point that at the end of *True Believers*, I took off my journalist hat and made an effort to influence and persuade the reader. The goal was to push the reader to get out there, read a book on the subject, and challenge their own way of thinking. *Stranger Than Fiction* is no different in that sense, but this time I will be presenting cases, ideas and theories that go beyond our perceived beliefs of what is reality, and why what isn't reality potentially *can* be.

> "*There's a field near the dream. I watched it grow with widest eyes. I watched us all reach out and read. Feel the strength as we touched the sky.*" – Angels & Airwaves

THIS CHAPTER WILL JUMP STRAIGHT into consciousness because the goal here is to punch the reader figurately in the face from the start with ideas, cases and theories that fly in the face of what we've come to know as true.

In the first book, *Punk rock and UFOs: Cryptozoology Meets Anarchy*, I theorized this idea that revolved around where do we go when we die? I will expand on that further, but here is a refresher.

A really interesting phenomenon when people have near-death experiences is that they all usually experience the same thing. You hear stories of them seeing a light or wormhole and hearing voices of those who've passed on talking to them. Sometimes they even claim to see or feel an alien presence. Research from neuroscience says that these are all hallucinations caused by various physiological and psychological factors. Some researchers say there is a transcendental explanation that cannot be easily explained by science.

Dimethyltryptamine (DMT) is a psychedelic compound with a structural analog of serotonin and melatonin. This has been used for healing or "divinatory" purposes in the past. It is speculated by some that endogenous DMT is produced in the human brain and is involved in certain psychological and neurological states. In the 1990s at the University of New Mexico, Dr. Rick Strassman advanced DMT research with a controversial hypothesis that a large amount of DMT is released from the pineal gland before death or during near-death experiences. The hypothesis says it was the cause of the near-death experience (NDE) phenomenon. Several of his test subjects reported audio or visual hallucinations. Several of his subjects tested also note "contact" with "other beings," alien-like, insect-like or reptilian in nature, in highly advanced technological environments where the subjects were "carried," "probed," "tested," "manipulated," "dismembered," "taught," "loved," and "raped" by these "beings." There is even a synthetic version produced that people take as a hallucinatory drug. Derived from plants, some take this as a psychedelic drug that produces powerful hallucinations.

What if our souls were a simple light force or consciousness? What if when we die, our "souls" go up past the ether into the atmosphere and find their way into another universe or a "heaven"? What if heaven is really another universe that can only be traveled to by our consciousness that contains our memories and cognition? All those "talking lights" seen in near-death experiences could be the souls of the deceased, and this universe or heaven is where our consciousness and souls all reunite.

Dr. Robert Lanza, who is considered the third most important current scientist by the NY Times, released a book called Biocentrism: How Life and Consciousness Are the Keys to Understanding the Nature of the Universe that says when the human body dies, life can still live on.

Lanza, known for his work previously in regenerative medicine, has been known for his extensive research that dealt with stem cells. Lanza was also involved with *physics, quantum mechanics and astrophysics. The professor took his knowledge of all the aforementioned theories to introduce biocentrism, which says that life and consciousness are fundamental to the workings of the universe. Biocentrism claims consciousness creates the material universe, rather than the universe creating consciousness, a huge middle finger to the established belief.*

Space and time are not objects, but instead parts of understanding. Many believe that no physical laws exist that would prohibit the existence of parallel worlds, a multiverse. Dr. Stuart Hameroff said in an article on spiritiscienceandmetaphysics.com that a near-death experience happens when the quantum information that inhabits the nervous system leaves the body and dissipates into the universe. Stuart and British physicist Sir Roger Penrose believe that our consciousness is stored in the microtubules of the brain cells. These areas of brain cells are the most important areas of quantum processing. When you die, the information (consciousness) is released from your body, meaning that your consciousness goes with it.

Now, what if, in some weird way, aliens served as a sort of guardian angels? Aliens have been believed to be visiting earth and human influence for thousands of years. The drawings of spaceships by Indian tribes, the

Egyptian pyramid carvings, the Mayan depictions, the stories of the "Gods" coming from the skies, which is a cross-cultural continental expansion of what is ideally the same sightings.

Factor in that many have described some of their alien abductions as an "angelic" experience and have claimed they've been visited their whole life by the same being, maybe some chosen few have alien guardians to watch over them. The ones with the recurring visitors cite recognizing the same aliens in their experiences and always having a calming presence when they are around.

Those who have an out-of-body experience or near-death experience see the lights, hear voices or aliens. We know there are thousands of galaxies that are out of our reach. Despite years of technological advancements, we aren't even close to attaining travel to those other galaxies, which surely host intelligent life. Interdimensional travel, intergalactic travel and time travel are all possible methods of transportation for aliens that human intelligence simply has yet to crack. The advanced alien science that allows them to travel to our planet, the method by which the UFOs operate, hover and use other out-of-this world abilities shows they are on a different playing field than we are, which is a scary thought if their intentions were cruel or colonial in nature. So, what if they are coming here to simply check up on us? Some *believe they've influenced and sequenced our DNA and assisted with our technological advances. Guardian angels?*

Some theorize that the aliens are visitors from the future and were possibly once of human form that inhabited this earth. An interesting thought considering if you believe in evolution and how physically they aren't too far off from a human form (eyes, nose, mouth, limbs, etc.).

Or how about this wacky one from yours truly: Aliens are what we eventually become as our consciousness and soul develops in the universe into something entirely different than what we were taught to believe. What if those alien visitors were really former relatives or ancestors from centuries ago, reincarnated of sorts, and now guarding over you like an angel? Crazy stuff, I know, but when we live in a universe with infinite possibilities, anything is possible.

So in a nutshell, aliens have always been there and have played some part in our development, and we cannot travel to these universes out of our realm of reach that aliens can, but what if someday our souls/consciousness could?

This idea that if our consciousness is our "soul," and this is how we are able to travel to other realms was interesting and "out there" at the time, but since then there's been more of a stake and shift to studying consciousness in and out of the UFO community. The "experiencer" is where the future of UFO studies is going, and could ultimately be the key to all of this.

An "experiencer" is defined as someone who has had an abduction experience or interaction. Abduction experiences have been reported all throughout history from people from all over the world, and all share commonalities. During this experience, the memories are usually blocked, or the "trauma" of it has hidden or shielded those memories deep into our own consciousness, and through regression and therapy we are able to access it.

But there may be another way to access these alternate universes, dimensions, and possibly planetary travel. If there was a gateway in our own minds that could open possibilities previously deemed impossible, then would that be an avenue worth exploring? The ancient Egyptians believed the "third eye" exists in all of us and holds powers that would be similar to those of a God. This third eye would be located on what we call the pineal gland. The idea of a third eye also appears in various religious references and is believed to hold visions, clairvoyance, the ability to observe chakras and auras, precognition, and out-of-body experiences. The third eye, also called the mind's eye in Taoism, is situated between the two physical eyes and expands up to the middle of the forehead when opened.

According to Strassman, DMT is released in large quantities from the pineal gland during death and near-death experiences. His breakthrough book on his DMT research, *DMT: The Spirit Molecule*, was an eye-opening scientific study that legitimized the psychedelic connection with altered states like birth, death and near-death, psychosis, and mystical/paranormal experiences. Volunteers would take DMT, and Strassman would record what they were experiencing in this altered state of consciousness, which brought beings, clowns, and other universes.

Strassman graduated from Stanford, obtained his medical degree from Albert Einstein College of Medicine of Yeshiva University, and was a tenured associate professor of psychiatry, and received the University of New Mexico General Clinical Research Center's Research Scientist Award. His intensive DMT studies draw instant synchronicity with paranormal and UFO/alien experiences. The ability of the pineal gland to naturally release DMT also aligns with what ancient Egyptians believed the powers of "the third eye" hold.

In his book, a quote that stood out was "one needs to travel without one's body." This idea goes with what I theorized that maybe in our current state we can't travel to other universes or galaxies but our consciousness could. This goes with Lanza's study of biocentrism and the idea that our consciousness lives on when we die. Is our consciousness really our soul?

"I think our consciousness is a function of our soul," Strassman said. "The medievalists believe there are several types of souls: a vegetative/plant soul mediating growth and reproduction; an animal soul mediating movement, instinct, and impulse; a human soul capable of self-reflection, self-consciousness, and connection to 'the divine.'"

With the feelings of euphoria and emotional reactions like those that are part of alien abduction phenomena with memories returning through regressions, meetings with "them" take place in nonmaterial worlds. Could it be that the DMT gateway is the way to access these "alien" encounters and different universes?

"To the extent that nondrug states resemble those brought on by DMT, it makes sense that endogenous DMT may play a role in mediating those experiences," Strassman said. "Assuming that those meetings take place in nonmaterial worlds, then there would need to be a way to access those worlds. However, that is a big assumption."

Gregory Shushan holds degrees in Religious Studies (University of Wales Lampeter), Research Methods for the Humanities, Egyptian Archaeology (University College London), and Eastern Mediterranean Archaeology (Birbeck College, University of London). He's also the author of *Conceptions of the Afterlife in Early Civilizations*.

"I'd have to question some underlying assumptions here – that NDEs are the 'same' as DMT experiences, and that the drug can be used to 'replicate' them," said Shushan. "Extraordinary claims require extraordinary evidence, and I'm not sure this has really been established. There are certainly similarities, but because both kinds of experience can vary so much between individuals, we need to tread lightly with straight one-to-one equations. Moving through darkness and emerging into light doesn't seem to be typical of DMT experiences, nor do meetings with deceased relatives. Feelings of being separated from the body are common, but I'm not sure that specific descriptions of leaving the body and seeing it below are. And more idiosyncratic imagery reported in DMT experiences – like giant insects or aliens – aren't really typical of NDEs, but to answer the question to the best of my ability: who knows? These experiences could be gateways to other worlds, or simply to other states of consciousness resulting from non-paranormal factors. In both cases – though especially in NDEs – they are profound and often life-changing, but this doesn't necessarily mean they're veridical."

So let's assume aliens have mastered the avenue into our consciousness, which would explain why they are able to come in the night and manipulate our senses and paralyze our minds and bodies. Do these other worlds and other beings exist just on the other side of our consciousness?

I always reference the book *Keepers Of The Garden* by Dolores Cannon, which most in UFO research would be wary to take seriously. If we are going to the highest degree of high strangeness, we can't rule out the weirdest of the weird. The book is a hypnosis case; through the hypnosis sessions with Cannon, a regular man named Phil is able to channel "imprints" of his past lives as many "alien" forms on earth and other parts of the universe. Through the channeling, Cannon is able to get information from a group of twelve all funneling information though Phil's subconscious. Before you quit reading now and dismiss this as nonsense, please allow yourself the ability to judge for yourself.

What is presented in this account seems to hold many possible answers to the world's mysterious ancient astronaut question. What was presented could be from the man's imagination and the psyche of a man under hypnosis, but there are a lot of coincidences to what we've all studied in terms of UFOs that connect a lot of the pieces of the puzzle in a manner that feels fluid.

Many beliefs of ancient astronaut theorists are confirmed here, but the juicy stuff comes with not IF aliens played a role in our development, but WHY. The ideas of different universe energies as imprints or "spirits" rather than what we would think of a physical human or alien manifestation has also aided in our development. In some cases, the "Gods" did descend on earth and show civilizations ways in astronomy, agriculture and medicine, but in other cases, and still to this day apparently, influence is felt via energy and ideas as inspiration. Through energy and imprints and channeling, many great minds have been influenced in terms of innovation from one plane of awareness to another. These are considered "gifts" rather than manipulation. Our body is just a vessel, and indeed our energies can live on. There is an important, universal interest in the well-being of earth, and spiritually we aren't quite there yet in terms of exploration and overall enlightenment.

Some important bullet points:

-Energy receivers remain on this planet in forms of pyramids, and in America (the Washington Monument is listed as one), and are an untapped resource on earth in terms of communication with the universe.

-The Akashic Record is also, indeed, a thing. Think

about the ideas of cosmic consciousness, zero-point energy and the Akashic Record, which is a belief that we can tap into the universe's "cloud" of knowledge. Zero-point energy, which was an Albert Einstein idea, is theoretically sound, and quantum waves exist, but are invisible. This idea makes up the Akashic Record: an intelligent informational field we can tap into through our consciousness, which you can compare to the current "cloud" and internet.

-Extraterrestrials are here in three ways: physical forms that visit us, those who live in secret bases on Earth, and spiritual beings undetected by human consciousness.

-There are many races of aliens just like there are many races of humans and breeds of animals.

-There are planets that host intelligent life that we can't perceive through sight.

-There are beings interdimensional in nature.

-Visitation cases and "tests" on humans are to experiment, take samples and prepare humanity for a potential move to another planet.

-There are some physical beings on our planet that are elusive and come from other energies that allow them to mostly go undetected (Bigfoot?).

-Tales of creationism do hold some accuracy.

-There's a reason these feelings and experiences are

held in our subconscious – described as "the information is buried beneath a highly disruptive cover of emotional trauma caused by the extreme warping of conscious reality."

What if these "crazy" ideas, which oddly coincide with biblical, mythological, fictional and scientific accounts and studies, hold the key to the truth? What if the ideas of energies and channeling these energies, as well as spiritual awakening where we tap into our own consciousness and energies, is what is needed to advance our understanding and move civilization forward?

A lot of trying to understand this phenomenon is like trying to put together an imperfect puzzle with thousands of pieces scattered all over the world. Will all the pieces fit? Will the final product make sense?

The cosmic ties to consciousness are slowly coming to the forefront of UFO culture, but at one point it was considered "New Age" nonsense to many who wanted UFO research to be mostly focused on sightings and applications of tech.

There are a lot of highly intelligent innovators, scientists and minds currently working to advance our technology and overall way of life, but there still is a brick wall hit when it comes to studying anomalous phenomena. First, our scientific processes, variables, and controls really can't measure a unit of study that we don't have a firm grasp on conceptually, or we can't control in the first place. UFOs act erratically and unpredictably, with the exception of some recorded flaps where UFOs show up in the same area over a certain amount of time.

There are some things that science simply can't explain, and that doesn't mean it should be discredited or considered a false narrative. For example, the young boy with no prior knowledge of World War 2 who draws fighter ships and mentions specific people who are real and served during the war. Where did these memories come from? How could he make it up if everything he says he knows was true? Is reincarnation just another reality that science can't solve?

"I think some part of the consciousness does survive bodily death; I'm not sure how you scientifically study that," said Jeffrey Kripal, professor of religious studies at Rice University. "I don't think the scientific method is the best way to know everything; it is good for some things, and bad for others. And I think that particular question could not be answered with the scientific method, but I think there is indirect evidence of both near-death experiences and children who remember previous lives, which are two different forms of the afterlife. One is a heavenly journey, and the other is a reincarnation story."

New York Times journalist and author Leslie Kean has written two groundbreaking books on the unexplained, *UFOs: Generals, Pilots, and Government Officials Go on the Record* and *Surviving Death: A Journalist Investigates Evidence For An Afterlife*. She acknowledges that the intersection between the larger UFO phenomenon and the exploration of consciousness are not that far apart. With many UFO-centric studies shifting to consciousness and afterlife, these once taboo topics now can be talked about in the same conversation.

"I think the question of the relationship between the UFO phenomenon and consciousness, the paranormal, and surviving death is a really fascinating one. Throughout my career, I have always thought of them as separate, and when I was working on *Surviving Death*, it was never linked to UFOs in my mind in any way. But now I've become more interested in finding that link," Kean said. "Other people have been working on the link for years and know way more about that than I do. I have come to see that there is some sort of overlap. We know that close encounters with UFOs impact people in paranormal ways, in terms of the abilities they develop or extraordinary experiences they have. For example, some experiencers report events that seem to combine seeing or interacting with deceased loved ones at the same time as alien beings. Or they develop paranormal abilities after encounters. There are a lot of interesting things going on between the edges of the two realities. I think it is really the next step, the future direction we need to go in – to find out how they connect and to integrate them and make that aspect acceptable to the mainstream. As we learn more about the nature of consciousness, we will learn more about the more paranormal aspects of the UFO phenomena, and vice versa. But I will continue my role as a journalist reporting on UFOs as unexplained physical phenomena that need to be taken seriously. In terms of reporting, we have to go one step at a time."

One thing Kean has noticed on the speaking circuit talking to younger college students is more of an acceptance to these topics, which in the long run can birth potential progress.

"I get a sense that there is a lot of interest in the younger generation, and they may be more open to these things since they haven't been as affected by the scientific status quo or restrictive materialistic views that we older people have been dealing with for decades," Kean said. "Younger people have a freshness and openness that is wonderful, as well as an intelligent curiosity, and they seem to accept these realities more easily. We need them to carry these studies forward."

Jon Sumple's documentary *Extraordinary: The Seeding* provides a detailed documentary that vividly describes the emotional trauma and long-term effects of abduction experiencers.

For many unaware of these themes, some of what is presented here may be hard to digest for the average viewer: reptilians, hybrids, and "otherness" that may sway some to dismiss the witness claims. Putting the experiencers at the forefront of this film is its bestselling point; you get their story told clear through their regressions, alleged abductions, pregnancies, sexual encounters, and consciousness/memory manipulation. Some of the common themes in the film include genetics, screen memories, reproduction, hybridization (hybrid children stories), and UFO phenomena. There are some good ideas here as one experiencer floats the theory that "consciousness is the power of the extraterrestrials."

If "consciousness is the power of the extraterrestrials," then how would we stand a chance if we don't even understand our own consciousness and humanity yet?

> "We demonize the unknown. We are afraid of it" – Witness of Another World

THROUGH THE ANNALS of Hollywood history there have been recurring themes of the invasion angles and "evil alien" agenda with films like *Independence Day* and *Mars Attacks!* There have also been lighthearted friendly alien movies like *E.T.* and *Paul* to counter the blow-'em-up popcorn flicks, which could all be part of the idea that Hollywood is in on the plan for a slow disclosure that includes pop culture. In these invasion films, despite being technologically disadvantaged in every way possible, humanity finds a way to win, but is this scenario even possible?

UFOs are embedded in pop culture with a rich history in film, TV and multimedia. In many sci-fi and horror films, the aliens are depicted as hostile and a threat, but Earth prevails in the end, which may not be accurate if an actual invasion were to ever happen. So why push the idea that we could survive? Is this to soften the blow that we cannot, or is it to stop mass hysteria if the truth ever came out?

"In most UFO invasion movies, humans – typically Americans – ultimately conquer the aliens because audiences tend to prefer their blockbusters to have upbeat endings, or at least for the good guys to come out on top," said Robbie Graham, author of *Silver Screen Saucers*. "Downbeat endings rarely equate to big bucks at the box office, so it's just good business sense to have the humans defeat the aliens. Over the past 20 years, an increasing number of alien invasion movies have been produced with the close co-operation of the Pentagon, which uses Hollywood whenever possible as a crucial conduit for political and militaristic messages. It's an arrangement in which various branches of the military provide film-makers with expensive hardware and onset advice ostensibly to help make the film more authentic and to help craft a positive image of America's armed forces, but this cozy, symbiotic arrangement also allows for the military to directly shape the content of the movies in which they are involved at every stage of the production, including the script-writing stage. As a result, in theory, the

Pentagon can insert all sorts of messages and ideas into Hollywood narratives, including those of UFO and alien invasion movies. We need only look at Pentagon-backed products like *Battleship* and the *Transformers* franchise to see how the US military envisions itself defeating an extraterrestrial foe, and naturally it's in the interests of the Pentagon to portray itself as being only too capable of defending America from any form of enemy, whether they're from Russia or Zeta Reticuli."

War Of The Worlds by H. G. Wells was a sci-fi novel about a Martian invasion and was adapted to American radio in 1938 by Orson Welles and caused real panic since many listeners believed it was an actual invasion. Films about UFOs (*E.T.* and *Close Encounters Of the Third Kind*, for example) were allegedly encouraged by the government to likely get the public's mind off actual sightings. The public was to be entertained, and this was "camouflage through limited disclosure" as alleged in the book *The Day After Roswell* by former Pentagon official Colonel Philip J. Corso, who worked closely on US government projects regarding reverse-engineering alien technology from the Roswell crash.

"In decades past, the Pentagon flatly refused to involve itself in any movie depicting UFOs or aliens," said Graham. "That's changed in recent years, however. Today, every branch of the US military is only too willing to lend support to UFOlogy-inspired entertain-ment products. Can this be interpreted as an indirect acknowledgment of UFO reality? Perhaps. Or perhaps the Pentagon realizes that, when it comes to boosting military recruitment, it can no longer afford to be picky about the genres with which it associates itself."

If you take into account the recent Pentagon UFO/Gimbal/*Nimitz* encounter mainstream UAP stories, you'll notice they all have credible military witnesses baffled by a craft that moved in a way that defied all physics. Our planes couldn't keep up, catch or replicate the technology of the Tic-Tac and the other UAPs caught on camera and eventually released in conjunction with To The Stars Academy Of Arts and Science. There have also been reports through time of UFOs showing up around nuclear plants, and common appearances among military bases with some reports of the UFOs being able to disable firepower and other human-invented tech. There's the claim that an Italian helicopter was attacked by UFOs, and various other historical interference from all over the world.

To The Stars Academy Of Arts and Science announced a partnership with the US Army to "advance TTSA's material and technology innovations" and "leverage developments in material science, space-time metric engineering, quantum physics, beamed energy propulsion, and active camouflage" according to the news release. By nature, the idea of foreign vehicles in our airspace is indeed a "threat" regardless of hostility levels, but what we see here is an example of a government institution attempting to ramp up their tech game.

"A lot of the time we think of this technology in small ways since we assume that the organisms we would be up against would simply be like another tribe of humans, but perhaps with more advanced technology or something," said Chris Cogwell, the former MUFON director of research/current host of the *Mad Scientist* podcast, who has a PhD in chemical engineering. "To me that is a woefully I view. It is possible an advanced organism wouldn't cognate in the same ways we do; they may not have similar sensory apparatus, may not have language in the same ways we do, may not have morals or a sense of individuality even. So why do we assume they would fight us in the same way? One of the first assumptions is that they would care to keep us around at all or feel any sort of limitations to the morality of war that we feel as societies now."

MarketWatch columnist Jurica Dujmović theorizes that the Tic-Tac UAP, which pilots couldn't get a lock on, is military made even though his article states if the UAP was manned by a human, "the vehicle would have to be equipped with technology capable of reducing the inertial mass of the object by generating gravity waves to reduce g-forces during acceleration."

Technology aside, common feelings among abduction experiencers are helplessness, paralysis, and memory loss. So if an advanced civilization was able to manipulate us, capture us, and contact us through consciousness, then do we have any way to combat this when most mainstream science will shy away from this idea? In *Silver Screen Saucers* Graham quotes theoretical physicist Michio Kaku, as he compared an alien-threat-versus-humans standoff as "Bambi meets Godzilla" with the earthlings being Bambi here. Kaku also offers the warning in the book of "we would present no military challenge to such an advanced civilization ... We would be a pushover for them; forget all the Hollywood movies."

All these factors make the idea of a Space Force (the upcoming branch of the military: land, water, air and now space) even more questionable since our technology may be considered primitive while standing next to advanced civilizations. Ultimately, if Hollywood could promote the idea we could win an "alien war," then maybe the public will be more at ease with "disclosure" one day.

One of the many components of To The Stars / To The Stars Academy Of Arts and Science includes the arts element, which plans to release movies and television shows. This plays into the idea of a slow disclosure through popular culture/fiction, but others will argue it is just another outlet to provide these extraordinary stories. Stan Spry is a producer and CEO of The Cartel, which will be releasing TTS properties that includes the TBS live action *Strange Times* series.

"Part of their mission is to spread information and knowledge and disclosure through entertainment," Spry said.

Sekret Machines is a three-part nonfictional series that To The Stars is releasing that coincides with a fictional book franchise of the same name that blends a fictional story with historical events and alleged UFO occurrences. Although many in academia understand throughout history fiction and nonfiction intercept each other, there is always the argument that blending the two "blurs the lines." How does having a fictional franchise in *Sekret Machines* and a nonfiction component position To The Stars Academy of Arts and Science to break that "blurring the line" criticism?

"We have 70 years of stigma and ambiguous actions taken by the US government to study and understand the UAP phenomenon," said Tom DeLonge of To The Stars in an exclusive statement for this book. "By using fiction to galvanize people to make sense of that history in combination with nonfiction to give them a deeper dive into the facts, we are able to marry the two tactics to serve as an impactful vehicle to bring people up to speed."

In studying humanities, media, pop culture, and mythology, we see that fictional works aren't determinantal to real scientific studies, as our technological advances today were very well science fiction years ago.

This certainly isn't the first case of Hollywood infusing nonfiction into fiction stories. UFOs and the unexplained are slowly becoming more acceptable in the mainstream cultural landscape, as a big part of the equation has been media, namely television.

There's been a significant change in the attitude towards the subjects of UFOs, the unexplained and other phenomena, as the stigmas are slowly starting to fade regarding these once taboo topics. The *New York Times* Pentagon UFO story, the Tic-Tac and Go-Fast To The Stars initiated video releases, and increased mainstream news coverage from print/online outlets like Politico, the *Washington Post*, and the *New York Times* have made it acceptable for the CNNs and Fox News Networks of the world to join in on the coverage. Often print media and reporting helps legitimize what makes it on broadcast.

In addition to the cable television coverage, which is

mostly driven by a "mainstream bias" of reporting hot topics, production studios have also jumped on board. There's been an influx of UFO and paranormal shows like CW's *Mysteries Decoded*, History's *Unidentified*, Travel Channel's *Alien Highway*, History's scripted series *Project Blue Book*, and others all popping up within the last two years. A second season of *Project Blue Book* has been ordered after outperforming most new shows last year in the 18–49 age demographic according to tvbythenumbers. History's *Ancient Aliens* is currently on season number 14. And for every show that we see on the subject, there have been plenty that have been plotted, pitched, piloted and in production that haven't made the air or are currently being produced.

"One of the big things at the networks regarding getting the green light for any 'unexplained' show, mystery or conspiracy is 'why now?'" said Gary Tarpinian, president of Burbank-based film and television production company MorningStar Entertainment. "Why are we producing this show now? If you can't answer that question successfully, you're not going to sell your show."

Unidentified played a large part in helping legitimize what the *New York Times* reported, and what To The Stars released regarding the Pentagon UAP videos like the Tic-Tac and the Go Fast. The show spent the majority of the six-episode first season documenting Luis Elizondo investigating the Tic-Tac case with the eyewitness military personnel on-camera testimonials, which was representative of what the media was giving serious coverage.

"This has happened in the past – in the early '50s; UFOs were national news, which prompted the creation of Project Blue Book – and now, the Navy's recent acknowledgment that UFOs exist is highly significant," said Steve Ascher, senior vice president of A+E Originals and one of the producers of *Unidentified*. "History's series *Unidentified: Inside America's UFO Investigation* played a major role in that moment. In our series, former military intelligence official and special agent in charge Elizondo, interviewed then Lieutenant Ryan Graves, whose words and shared experiences with UAPs as an active-duty Navy pilot got us here. Graves was the first USS *Theodore Roosevelt* pilot to come forward to identify the two videos of UAPs, setting the internet ablaze. Our series has presented information that has never been disclosed before, and sparked conversations that have initiated true government action regarding this topic. We are thrilled to have played a small part in making history."

Television reaches 90 percent of the US population,

and the broadcast industry is a $150 billion per annum enterprise according to *Business Insider*. Due to the cheaper production costs and consistent ratings, reality-based TV is at the forefront of the industry, and these informative, yet slickly produced paranormal/UFO-based shows are a part of the machine. The machine is also in direct competition with each other for new and groundbreaking content.

"When developing and selling a UFO-based series or documentary, there are a couple of things that most companies are looking for; what is different about this story that hasn't been told before?" said Chase Jeffrey, development producer at Megalomaniac Inc. "Is there new access we have yet to see? Is this newly declassified information? Did some event recently happen that had made this topic relevant? Did a public figure recently make a statement or get involved in the movement? In our industry we are always the 'first' to tell the story – and if it's an iconic story that has already been told, then what's the new angle? We always seek to find new ways to break through the clutter in hopefully a new fresh take ... It's a very competitive industry."

Director Seth Breedlove and his Small Town Monsters productions have garnered the attention of Hollywood with their DIY films, and the positive press they have received. For the time being, Small Town

Monsters remains independent of any TV contracts. Breedlove argues the homogeneous nature of some of these cryptozoology shows have clouded the creativity of the production companies.

"My experiences with television production regarding the paranormal have been the same story told over and over," Breedlove said. "Basically, they all want to copy whatever the next guy is doing, regardless of whether or not the model is a success. There is very little, if any, actual knowledge of the phenomenon they're making a show about, and all the focus is on personality and how to portray that personality onscreen. I've been pitched at least a dozen versions of a show called 'Small Town Monsters' by various production houses that all boil down to 'you and so and so are out in the woods hunting for monsters.' That's not to say that all TV development people think this way, as I believe that may not be the case, but the ones I've dealt with have all been that same sort of experience."

Despite having TV opportunities, Greg Newkirk and his Planet Weird media company skipped Hollywood and went straight to the streaming services like YouTube and Amazon with its paranormal series *Hellier*. *Hellier* received positive press and a different feel than the usual TV paranormal investigation series by avoiding the status quo network shows followed.

"When we made the decision to create, produce, edit, and distribute *Hellier* on our own, it was because we couldn't bear the thought of changing it to fit the formula," Newkirk said. "We'd created a very real look at the inherent messiness of researching paranormal phenomena, complete with dead ends, high concepts, and the incorporation of elements like UFOs, cryptozoology, ghosts, witchcraft – and we didn't segregate them – a concept that makes networks nervous. There's a real-life narrative that pushes our show through every episode. To boot, our director, Karl Pfeiffer, shot the series with an anamorphic cinema lens and color graded it like a film; something no one has ever done for a paranormal television show. In the early meetings we'd had with network folks, they told us we couldn't do any of those things."

While many of the programs that cover UFOs or cryptozoology have the same themes and experts, studios are trying to break away from that trend by introducing new voices and concepts. A lot of these concepts don't ever make it to TV at times.

"I think that what we look for in paranormal/UFO stories is primarily driven by a new approach or way to explore them," said Alex Hoff, vice president of Hoff

Productions. "It could be new footage, unique accounts and experts, and even just a new story-telling method. Like most shows, there's a finite amount of these stories out there, so you have to find new and engaging ways to tell them to people who oftentimes consider themselves experts and have a pretty good BS detector."

Ultimately, these shows exist because there is a demand and an interest from the audience. These shows are being taken seriously in tone, which could be a reflection of how the news media is now treating these cases. If the show can make an impact on the audience, it is more likely to come to fruition and be successful.

"Another thing the network brings up is 'Why should our viewers care about this story?'" said Tarpinian. "You also need to answer that question, too – or you're not going to get an order. Though there is more than one answer to this question (depending on the topic and who the viewer is), I believe that if we show the viewer how the story may affect or impact them, then you are going to get their attention, and you are in business."

When the *New York Times* published the piece titled "Glowing Auras and 'Black Money': The Pentagon's Mysterious UFO Program" on Dec. 16, 2017, it was a paradigm shift in the way the public and mainstream press outlets presented stories on UFOs.

Since that initial story by veteran journalists Helene Cooper, Ralph Blumenthal and Leslie Kean, a lot has happened in bringing this topic away from the fringe. When the first *Times* article hit, punkrockandUFOs-.com went behind the scenes of the story with Blumenthal, and he discussed how they considered Elizondo credible and wanted to pursue the story.

"The good news is the *Times* is receptive to this story," Blumenthal said. "They have to be done in a *New York Times* way where everything is on the record. In this case we did, we were able to talk to several pilots without using names. Everyone else was on record. The *Times* takes the subject very seriously with its editing standards, which is why the stories have gotten such credible resonance. When people see it in the *Times*, it's not fake news, (and) it's not speculation. We are very careful to stick to what we know, and lay out what we don't know. The editing process is very rigorous, and it is not easy for reporters like us to get these stories through because we are held to a very high standard. Given the subject, I would say properly so."

The *New York Times* isn't the only journalism giant taking this seriously all of a sudden. *Politico*, the *Washington Post*, *The Hill*, *New York Mag*, and other large media companies are following suit with similar stories, which makes the journalists at the *Times* prideful from what started as the Pentagon UFO story to now.

"We love to see it lighting up the internet," Blumenthal said. "It is the most e-mailed, most watched and most everything'd story of the day. It is really dominating *Times* coverage, so it is a part of pride because we do think the subject is important, and we love to see people follow us since sometimes we need to follow other people, but in this case I think we have set the pace."

With the new *NYT* story "'Wow, What Is That?' Navy Pilots Report Unexplained Flying Objects" that dropped May 26, Blumenthal, Kean and Cooper returned to feature two new witness testimonials from military officials with the biggest takeaway being that these baffling UFOs appeared almost daily from the summer of 2014 to 2015 off the East Coast. The article features quotes from Lieutenant Ryan Graves (a ten-year Navy vet) who reported his sightings to the Pentagon and Congress, Lieutenant Danny Accoin, and three others (who asked not to be named), who spoke on record to the *Times* on what they saw. Graves and Accoin are mentioned in the story and also appeared on *History's Unidentified: Inside America's UFO Investiga-*

tion. The sightings reported here seem to mimic the descriptions of the 2004 *Nimitz* UAP case.

The case has helped launch UFOs/UAPs into the mainstream, and the story is being taken seriously.

"The primary reason I began reporting on UFOs was that I felt there were legitimate sightings that appeared to be genuine mysteries," said veteran UFO journalist Alejandro Rojas from Open Minds TV and DenOfGeek.com. "I was especially interested in sightings by law enforcement or military personnel that government researchers like Project Blue Book could not resolve. Despite these cases, for decades, mass media made fun of the topic and added jokes to it even while reporting on credible incidents. Now that the Pentagon has admitted they investigated UFOs, and the Navy has shared some of their best cases and have said they take sightings by their personnel seriously, UFOs are a more credible topic. Unfortunately, it has not stopped the conspiracy-minded from trying to fit these new revelations into their predetermined theories. However, I am excited about the future of UFO research, especially as conventional organizations and government agencies begin to publicly investigate the topic or share the research they have already done behind the scenes."

> "*Sometimes truth is stranger than fiction.*" –
> *Greg Graffin (Bad Religion)*

KEVIN DAY WAS the air intercept controller (AIC) aboard the *Princeton* during the now highly publicized *Nimitz* UAP incident. Like many people who've experienced UFO phenomena, Day's life changed that day, and like many, it compelled him to tell his story in one way or another.

Day hid the truth "in plain sight" in his 2008 fictional story "The See'r" that mirrored his experience witnessing an aerial phenomenon unlike he's ever seen – the Tic-Tac encounter. His short stories comprised his books *Sailor's Anthology Book I* and *Book II*, but like many confused with the truth presented to them, he told his story through fiction.

"The weirdness of the Tic-Tac encounters left me struggling to make sense of what had happened," Day said. "In the four years following those events, I made a couple of attempts to tell my story to those closest to me. The result of my effort was likely predictable. My story was met with a sort of kind skepticism. The combination of not knowing what the *Nimitz* Strike Group had encountered, and reactions I was receiving from the few people who had heard my story, I decided to write about the incident at sea in a way that accomplished two things for me. 'The See'r' preserved the larger story – the facts concerning what took place in the Combat Information Center (CIC) aboard the *Princeton* in the case the Tic-Tac encounters ever did come to light – and writing it as fiction provided me a way of attempting to describe the very nature of what we had encountered. To capture the weirdness of it. And although 'The See'r' is certainly fantastical, so too was the actual event. In fact, it is entirely possible the actual truth will turn out to be far, far stranger than my fictional story line."

Since his experience, the protagonist in his stories mirrored his real-life experiences during and post-*Nimitz*. He was visited by what he described as "spooks" and claims to have advanced cognition. Day's consciousness also shifted in his dreams. After the UAPs he witnessed, he began to be haunted in his sleep by eschatological and militaristic themes.

"The dreams I began to have in 2008 can be loosely described as eschatological: world-wide disasters, comets causing tsunamis, epic floods, earthquakes, plane crashes, (and) end-of-the-world scenarios," Day said. "I remembered the 'nightmares' the next day and those dream-memories would trigger acute anxiety, which I experience daily even now many years later. Sometimes the anxiety becomes so intense that I flashback – remembering the dream surfaces other real memories – and I suddenly 'zone out' for a short time. It is sometimes so intense that other people present have asked if I am OK, which I am after the extremely unpleasant episodes are over. If not for the anxiety, perhaps the dreams themselves would not bother me so much. They're just dreams."

In his story, he mentions an extremely powerful energy manifested in consciousness that could be good and bad, which corroborates with what many experiencers and believers in metaphysics mention.

"It has been observed that close encounters (CE) with unknown aerial phenomenon (UAP) often causes notable post-effects in human observers," Day said.

Day references a 2003 paper by Jacques Valle and Eric Davis.

. . .

In 2003, Dr. Jacques F. Vallee and Dr. Eric W. Davis wrote a peer-reviewed paper entitled "Incommensurability, Orthodoxy and the Physics of High Strangeness: A 6-layer Model for Anomalous Phenomena." In their paper they describe certain physical and 'anti-physical' manifestations of UAP encounters.

The paper's abstract explains, "The main argument presented in this paper is that the continuing study of unidentified aerial phenomena ('UAP') may offer an existence theorem for new models of physical reality. The current SETI paradigm and its 'assumption of mediocrity' place restrictions on forms of non-human intelligence that may be researched. A similar bias exists in the Ufologists' often-stated hypothesis that UAP, if real, must represent space visitors. Observing that both models are biased by anthropomorphism, the authors attempt to clarify the issues surrounding 'high strangeness' observations by distinguishing six layers of information that can be derived from UAP events, namely (1) physical manifestations, (2) anti-physical effects, (3) psychological factors, (4) physiological factors, (5) psychic effects and (6) cultural effects. In a further step they propose a framework for scientific analysis of unidentified aerial phenomena that takes into account the incommensura-

bility problem." In my case, I seem to have been affected most strongly by Layer V, psychic effects. Vallee-Davis describe these effects further "... impressions of communication without a direct sensory channel ... poltergeist phenomena ... motions and sounds without a specific cause, outside the observed presence of a UAP ... levitation of the witness or of objects and animals in the vicinity ... maneuvers of a UAP appearing to anticipate the witness' thoughts ... premonitory dreams or visions ... personality changes promoting unusual abilities in the witness ... healing ..." (Vallee-Davis, 2003, pg.8, para.3).

Sean Cahill is the charismatic key witness of the famed *Nimitz* encounter, and his story and continued exploration played out in History's hit *Unidentified*. Cahill is currently a retired US Navy chief master-at arms and is spending his civilian life as an investigative filmmaker and meditation facilitator.

Cahill is one of many highly trained former military personnel who have been outspoken about what happened with the Tic-Tac UFO/*Nimitz* incident, as well overall "disclosure" when it comes to the conversation of UFOs, what he witnessed, and frustrations associated with the stigma that surrounds the phenomena. Cahill is raw, passionate, thoughtful, and determined to do his part to serve the greater good, which is what he signed on for as a member of the Navy.

"You could say that I was a UFO nerd my whole life up until that point," Cahill said. "My experience in 2004 had an effect opposite to what many other people report. Rather than an awakening, I went back to sleep. I had already been a curious and spiritually skeptical person, but that went hand in hand with an open mind that saw an immense universe of possibility. I thought that the chain of command's response and seeming uncaring attitude spelled out that what we had seen was a known asset to the bigger machine. At the time I thought that meant our own technology. That frankly pissed me off. It was utter bullshit that not a single head turned above the position I thought that it was because it was above my pay grade. I think in the end the problem wasn't pay grade; it was clearance. What I lacked was the clearance to know that this was a real and embarrassing problem that we did not want to be shouted from the rooftops, pardon the pun, but it was still unidentified. I think I had it right and wrong at the same time, I think that duality is key to getting to the bottom of this ... I think the future is a shared conversation. If we the public give this subject back to the intelligence folks and the military-industrial complex to manage on their own in the dark, then it will be managed completely by fear. I think there is an unknown here that we must defend against until it shows itself to be a benefactor we can trust in a way we can verify. The people who came before did the best they could with what they had; our culture has a long history of shooting the messenger and shaming the pioneers as they sail away. That would be a mistake to continue. We need courageous acts in the neighborhood and at the dinner table. Ask Grandma what she has seen, and listen with an open heart. If we give our elders a chance to speak in safety and love, then we often find our own voice is stronger and the common ground that bridges the years."

Being ex-Navy, Cahill is also aware of the history of UFOs (or USOs – unidentified submersible objects) and the relation of sightings to military installations.

"I think a nuanced look at history shows that the remnants of humanity rallied around the new idea of agriculture after the last cataclysm, and that led to different population centers going forward; however, humanity has always settled around resource-rich areas, of which the sea is a major source," Cahill said. "It has served as a medium of rapid travel and trade among far-reaching cultures. Today, while it may only represent 16 percent of our global protein intake, the sea still covers over three-quarters of the surface of the Earth. The global economy would dissolve overnight without trade from the sea. Further, when you overlay our shipping lanes, commercial and military flight patterns, and training areas, we only touch upon a small fraction of our planet with human senses. Those same human eyes have to pour all the recorded data from satellites and Earth-based sensors through a sieve of computer algorithms and nearsighted analysts to find that one pixel moved. This is a water planet; we live in the tiniest membrane between uninhabitable zones. We do not control the hydrosphere or lithosphere or low earth atmosphere by any means. We inhabit a thin gray area of around 24 kilometers up and only 10 down. We control even less. Here's the short answer. We are impulsive jerks. We utilize nuclear reactors (mobile controlled meltdowns) for the power plants of our most destructive weapons systems. Some of those weapons systems can deploy multiple nuclear warheads to any spot on the globe in minutes. Whether we clandestinely and against deteriorating treaties keep a hot-load in the chamber, I don't know. I think that we can all agree that anyone willing to spend their resources observing and possibly interacting with us at this distance, or willing to expend the energy to transcend time and dimensions, would probably also be concerned with thousands of deteriorating high-yield nuclear weapons controlled by impulsive monkeys who, by the way, have no plan to dispose of any of these weapons or their waste products and have already used them twice in war. I think we can start with that and build from there."

In studying flaps of UFO sightings don't rule out anything.

Former NASA research scientist and current associate professor of physics at the University of Albany Kevin H. Knuth drafted an interesting hypothesis that ties the *Nimitz* UAP incident to the marine life in the area. The nine-page study even goes on to state that the UAPs could be monitoring the migration of whales, and even mentions a recent UAP sighting off the coast of Oregon similar to the Tic-Tac.

"The thought was triggered by Kevin Day's description of the UAPs as acting as if they were migrating," Knuth said. "They also seemed to act like they were looking for something by slowly tracking south at 100 knots or so at 28,000 feet altitude."

The study notes that the UAPs were around an unidentified submerged object in the ocean.

From the study:

Fravor Encounter On Nov. 14, 2004, Cdr. David Fravor, flying an F/A-18F Super Hornet fighter, was directed to intercept one of these UAPs. On reaching merge plot, Fravor and his Wingman, did not immediately notice the Tic-Tac UAP. Instead, it was observed that there was a large turbulent oval area of waves breaking as if over the surface of a submerged object (10–15 feet below the water) about the size of a Boeing 737 jet airplane. LCdr. Slaight thought that the object might be a submarine, which was later dismissed due to the fact that there were no submarines in the area at that time. The Tic-Tac UAP was then observed to be hovering with erratic motion over the churning water. An encounter with the Tic-Tac UAP ensued. After the encounter the submerged object was no longer present. It was never determined what this object was (Powell et al., 2019). Clearly, the UAP was interested in the submerged object. It is possible that this object and others like it are the reason that the UAPs were in the area.

The hypothesis also looks at various other factors like hydrography, chemical studies, underwater topography and biology. The gray whales are known to migrate south at that time of year from Alaska down to Mexico (past Guadalupe) and pass through the Channel Islands, which are where San Clemente and Santa Catalina are located. The area is also located close to military bases, which is another common hotspot for UFO activity and is another avenue to explore why the Tic-Tac UFOs were so prevalent in this specific region.

Star Trek IV: The Voyage Home also included whales and spaceships.

Patrick Hughes wasn't a direct witness to the *Nimitz* UFO encounter, but as an avionics technician during the incident, he was able to detail specifics on how it was viewed behind the scenes. He was a 2nd class petty officer (AT2)/an avionics technician with VAW-117 (the world-famous Wallbanger) at the time of the incident. He left the Navy in December of 2010 after 11 years and worked for Northrup-Grumman as a Tech Rep on the E-2C/D Hawkeyes for several years. Hughes left the Navy as a 1st class petty officer, but still remembers the historic Tic-Tac UAP sighting and is a part of the *Nimitz Encounters* short film.

"I never directly saw the Tic-Tac object; I've only seen the short clip most of the public has seen," Hughes said. "As I've stated in Dave Beaty's *Nimitz Encounters,* I dealt with behind-the-scenes stuff. The E-2C Hawkeye was my plane, and with its suite of sensors, it's pretty unstoppable. Most of what the sensors see is recorded onto hard drives. After a specific flight, (Dave) Fravor's flight of my memory is correct (but definitely the same day and near the same time) two Air Force officers accompanied my commanding officer and took the hard drives from the flight and never gave them back. Specifically, I believe they were after our CEC data. Can't really elaborate on why because much of CEC is classified. You can google CEC to get a basic idea. I can also relay things witnessed by one of my friends/coworkers who was airborne in the Hawkeye at the time. 'Roger's' portion of the story. He signed an NDA and that is why he isn't coming forward other than brief confirmation."

Hughes wasn't asked to sign an NDA. He was also not interviewed about this subject at the time of this interview outside Beaty's film, while other colleagues like Fravor, Beaty and Day have been featured.

"I never signed an NDA regarding this event, nor was I ever talked to about or asked about it," Hughes said. "The closest thing to any of that was our chief, who said shut our mouths when 'Roger' started sharing what happened. It boils down to compartmentalized information. All I did was hand over classified material to my CO. Something that's not exactly rare. It's amusing how the NDA topic is being tossed around ... I've signed NDAs for programs I worked with that my own chain of command wasn't aware of because they weren't read into the programs ... It's an insanely credible event. Everyone wants their part of the pie. I just want to know what it is."

> "*In the history of monotheism, which includes the Bible, the Bible is filled with what we would call paranormal stories.*" — *Professor Jeffrey Kripal*

THERE'S BEEN a synergy in told tales throughout time that blend fact and fiction. Real life influences fiction, and in some cases, fiction influences real life, as modern advances in science today were the fictional thoughts of science fiction writers from yesterday.

Jeffrey Kripal is a professor of philosophy and religious thought at the prestigious Rice University. He's also an author who has written books about the paranormal, UFO phenomena, science fiction/superheroes, religion and mysticism, and co-authored a book with fellow famous experiencer Whitley Strieber. Kripal is the type of academic you wish you had as a professor in college; he's engaging, thought-provoking and lectures and writes about topics that are right in the wheelhouse for those into the unexplained. His office is adorned with various books about his studies (UFOs, religion, mythology), large busts of iconic comic characters like Iron Man and Magneto, and a library that lands right in the middle of theism and pop culture. His book *Mutants and Mystics* is a massive study that intertwines pop culture and mythology by breaking down and dissecting some of pop culture's and comics' most iconic figures like Superman and Spider-Man from their origins, to their creators, and all the other connections that exist. Kripal believes religious texts are rich in what we'd call "paranormal stories."

"If you listen to people today talk about their spiritual experiences, they don't talk about going to heaven and seeing saints, they talk about entering another dimension, or spiritual evolution, or they don't talk about Jesus coming on a cloud with trumpets, they talk about UFOs coming from the future," said Kripal. "These are all religious mythos, but they are shifting dramatically as our understanding of the natural world changes. And I don't mean myth in the sense of falsehood; these are the deepest stories we can tell about ourselves and they carry truths that can't be communicated in any other way. ... My own conviction is that these types of anomalous experiences are as common as water. They are everywhere. And there is a very thin cultural shaming that holds it down and tricks us into thinking they are rare or crazy. For me, it is for people to start talking about the experiences no matter what they are; artists , musicians and people who create culture – then other people will jump in and other people will join in and they are no longer abnormal or shamed."

As an esoteric academic who has studied the world's history through a lens that focuses on the obvious and the underlying elements of historical narratives, Peter Levenda's expertise on religion, the occult, mythology, and UFO phenomena makes him one of the unexplained's most interesting assets. It would be lazy to pigeonhole Levenda as an occultist like some have, as his work as an author transcends just magic and mysteries.

Tom DeLonge enlisted Levenda to take the lead on writing the first of a nonfiction Sekret Machines trilogy for To The Stars. The first book, *Sekret Machines: GODS*, looked at the ancient astronaut hypothesis through a "cargo cult" perspective that took a deeper dive into the work of Zecharia Sitchin and Erich Von

Daniken. Levenda's approach ultimately helps legitimize the general scope of the idea that UFO phenomena are real and have been well documented through our earliest history until now.

"The instincts of Zecharia Sitchin and Erich von Daniken should be considered apart from the evidence they provide and the extreme claims they make," Levenda said. "Their instincts that there was some kind of contact in ancient times between *Homo sapiens* and some other race or beings should be considered seriously and calmly, but when every anomaly is associated with 'aliens,' then we are on very shaky ground ... I believe it is better to focus on the fact that *Homo sapiens* was around for hundreds of thousands of years and then just started to develop what we think of as civilization in the last twenty thousand years or less, all over the world, and with remarkably similar ideas concerning the gods, the heavens, life after death, etc. I write about this contact as a kind of traumatic event that imprinted itself on human beings: an event that we relive in every culture in an effort to find some sense to it, some positive way of dealing with the knowledge that we are not alone and indeed may not be the smartest kids on the block. Hence the 'cargo cult' theory. In my case, I have expanded the theory to include not only religion but science as well, and the organizing principle behind societies that focus on the heavens. In ancient times, those for which we have written records, such as Egypt and Sumer, religion and science were inseparable. They were part of a single worldview, a unitary approach to understanding reality. In modern times, we tend to put religion and science in separate boxes, and suspect that science disproves religion. But the goal of our most advanced scientific enterprises is the same as the goal of ancient Egyptian religion. For example: to travel to the stars. In Siberia, shamans climbed trees and then disappeared into space to travel to the heavens and speak with spiritual forces that live there. Humans built ziggurats in Sumer and in Mexico and Asia with the same purpose in mind. While religion and science became estranged from each other in the last millennium, they are working back towards each other again with some of the most advanced scientific concepts looking suspiciously similar to the understanding of mystics from India, Tibet, China, Africa, the Middle East and Europe. We are all still trying to get to the stars; that goal has not changed in ten thousand years. All we can ask is: why?"

These nonfiction books are paired with a fictional series of the same name co-authored by DeLonge and A. J. Hartley. *Sekret Machines: Man* by Levenda and

DeLonge is a good representation of what To The Stars Academy of Arts and Science is looking to ultimately achieve, which is to inspire a new generation to reevaluate and reimagine science, society and humanity.

A few interesting points:

-The experiencer experiences a response that is "spiritually elevated while emotionally devastated."

-Levenda makes the connection that aliens appear to have no knees, as it aligns with spiritual beliefs of deities and angels also being without knees.

-Aliens act more like "machinery than biology."

-The line of aliens are "not all powerful, they are just powerful in ways we are not," which plays into the idea of a symbiotic relationship.

-Humans may be evolving into a more consciousness-elevated being, and the existence of the phenomena is helping rewire our brain, perceptions and consciousness.

"Religion, mythology, and recorded history go hand in hand ... but so do religion and science/technology," Levenda said. "We ignore the relationship between science and what we call 'religion' at our peril. The irony is that there are religious fanatics in control of weapons of mass destruction: weapons developed by scientists and technicians, many of whom are atheists. We ignore the relationship between politics and religion at our peril, too. America famously has enshrined separation of church and state in its Constitution, but that is mostly theoretical when it comes to religiously motivated political leaders, elected officials submarining their religious views into public policy. Jimmy Carter was a benign example of that: his religious views were (and are) founded on high moral principles. Ronald Reagan, on the other hand, belonged to an apocalyptic denomination 'the Disciples of Christ' that saw the world in terms of the End Times and, incidentally, was the same denomination that gave us Jim Jones of Jonestown fame. Reagan famously framed the Soviet Union as 'the evil empire.' The atheists have presented the fanatics with the tools they need to slaughter millions and keep the globe in a perpetual state of violent chaos, but since they don't believe in God or religion, they wash their hands of any responsibility. Like all religious fanatics, atheists believe that if just everyone were like them, there would be no more wars and no injustice in the world. Whether one believes in religion or not, in god or gods, in angels or demons, it is only logical to accept that billions of our fellow human beings do hold these and many other unscientific beliefs and act on them. We need to get back to a reappraisal of religion, what it is, how it began, what started it, the consciousness aspects of it, the social and cultural aspects; all of it, from every discipline, and when we do, I predict we will come to a corresponding understanding of science and reality itself, and that will open our eyes to understanding the phenomenon."

Many of these components of unexplained phenomena all somehow meet in the middle. All elements of the unexplained deserve attention and seriousness despite what may sound ludicrous to some.

"Magic, mysticism, occultism, esotericism all have contributions to make to this study since they occupy the middle ground between religion and science," said Levenda. "Esotericism in contradiction to how it is normally understood, especially by its opponents, is preeminently pragmatic. It's a technology, a system, a method for attaining altered states of consciousness for the specific purpose of bringing something back that is useful. It just may be that what the magicians have found, and have found many times over the centuries, is what I call the 'Alien in the Panopticon.' (Obligatory Foucault reference.) We don't control when 'they' show up. We can't predict when or where. In terms of the phenomenon, national sovereignty is a joke. They're there. Whatever they are, and wherever they come from, they're there. They're in the central column of the reality Panopticon, and we are in the cells never knowing when we are being watched. Or why?

Wouldn't it be funny if the cell doors have always been unlocked?"

In the last chapter of *Gods*, Levenda mentions how when humans had a God among them (Jesus), they killed him. Do we see this idea play out in popular culture? Absolutely. How is Superman depicted in Zack Synder's film *Batman v Superman: Dawn of Justice*?

The film was decisive for its darker tone, but did play on themes of religion, mythology, and politics, which was fueled by the idea that if Superman existed in the "real world" today, how would he be viewed? Historically, people who are different are not viewed or accepted right away in many societies and cultures, especially America, where the seeds of division, classism, racism, and genocide are part of the nation's birth.

So, if a "Superman" came down from the sky today, would he be viewed as a Jesus-like savior or an existential threat? Would he act unilaterally, or would he serve a greater good? Would Superman be America's biggest weapon of mass destruction or defense? If an exceptional being with abilities that would make it basically undefeatable walked among us, then how would humanity respond and counter it?

Now if the idea of superior beings and metahumans exist, then how does humanity offset that? In the DC universe you have Batman, and when he encounters adversaries with abilities or when he even goes against Superman, you have the basic man versus God scenario. Batman doesn't have any special abilities. After witnessing the random murder of his parents, Bruce Wayne dedicated his life to crime fighting and injustice. Wayne's intellect, training, resources, financial privileges, physical shape, and drive are his "powers."

So how does a mere mortal stand up to a God? In Batman's case he knows Superman's weakness is kryptonite, and his humanity, which Batman is able to exploit because at the heart of it Clark Kent is just as "human" as the rest of us.

In *Batman: Hush* the line of "Even more than kryptonite, he's got one big weakness. Deep down, Clark is essentially a good person, and deep down I'm not." Batman's commentary here on why he holds the edge against Superman despite being physically outmatched is that humanity is intrinsically flawed.

Superman, as a God-like figure, holds the moral authority despite being an "alien" and different from us all. Is this any different than how an outsider with strange abilities and beliefs was viewed when Jesus was persecuted?

There are glaring similarities between Superman and Jesus. The story of Jesus appears in various religions under different names – the first dated back to the Egyptian story of the God Horus. On its surface, a heavenly father sends his only son down to save the earth. The son is raised by two mortals (Mary and Joseph / Jonathan and Martha Kent). Once he discovers his powers, he uses those powers to help those around him, which poses a threat to the powerful institutions. He eventually dies and is resurrected. So we have the same story told in many religions, and then it repeats itself via influence in popular culture.

Superman wears an S on his chest, and in the DC Universe films the S represents hope. The S is inside a triangle, and that shape is the symbol for the holy trinity: the father, the son, and the holy spirit. The triangle shape shows up a lot in UFO lore and mythology; triangle-shaped crafts, pyramids and monuments built in that shape, and reported sightings of aliens wearing a tight suit with a triangle on the chest. Hmmm, does that sound like someone?

While on the subject of a trinity, a DC Comics series that includes Batman, Wonder Woman and Superman is called *Trinity*. It's an interesting name for the group since you could mirror this idea to the holy trinity. Batman could represent as a mortal human the "son," Superman with his extraterrestrial abilities and savior-like mentality the "holy spirit," and Wonder

Woman with her Amazonian Goddess stature as the "father."

We tend to view Superman as not just a hero, but the standard-bearer for morality, even though he's an alien. Religious followers and theologians hold God and Gods to a higher standard than humanity.

"Superman presents himself to us as first and foremost a friend, a helping hand," said Zach Moore, host of the *Always Hold On To Smallville* podcast and Superman historian. "Because of his powers and abilities, he is seen by most as a savior, but to others he is seen as the most dangerous being on the planet: an invader, a false prophet, or even a false god. Some, like his arch nemesis Lex Luthor, even frame his presence and influence as a hindrance to our society, weakening us by being an outsider solving our problems and from a certain point of view 'taking our jobs.' Ultimately, it is Superman's actions that win over the public to his side even in the face of adversity, his relentless pursuit of 'truth, justice and the American way' and commitment to defend the innocent at any cost proves that he is what he says he is: a friend. Humanity is prone to fear the unknown. The other, anything or anyone they do not understand, but if they can put away that fear and judge someone for what they say and do as opposed to where they may come from or any preconceptions or preexisting prejudices, that's what will take us closer to the world of tomorrow. Superman is the light that shows the way. He is the ultimate immigrant story, having come to this planet from far beyond but survived and thrived here because of the care and kindness of others."

We already know there are beings on this planet that hold extraordinary abilities that are not human. With all the talk of space exploration and search for alien life, often the ocean and all its wonders and mysteries get ignored.

Euhemerus, the Greek mythographer from the late-fourth-century BC time period, attempted to rationalize myths. Euhemerus was one of the first, but certainly not the last, to philosophize the idea of finding truth in myth. He suggested myths may have originated in real historical events, and mythical characters may have been real people. Euhemerism stems from Euhemerus' beliefs and is considered the "historical theory" of mythology.

Andrew Lang was an author, folklorist, anthropology field contributor and served as the president of the Society of Psychical Research in 1911. Lang wrote fictional books on fairies and wrote widely on topics of folklore, myth, and religion. One of his beliefs was that we can explain everything by a recourse to "common sense," but supernatural beliefs might originate from real experiences and aren't as "irrational" as paranormal claims are sometimes made to be.

Comparative religions, near-death experiences, and cultural connections are just some of the factors that weigh into "paranormal" research. Dr. Gregory Shushan is considered one of the leading authorities on near-death experiences and the afterlife through history and culture. He is an honorary research fellow at the Religious Experience Research Centre, University of Wales Trinity Saint David, and has authored the books *Near-Death Experience in Indigenous Religions* and *Conceptions of the Afterlife in Early Civilizations*.

There is a cross-cultural commonality between NDEs, and there are also cross-cultural paranormal and UFO abduction phenomena. Why do so many "paranormal" occurrences happen cross-culturally?

"At risk of seeming overly pedantic, I would question the assumptions behind the word 'happen,' because it seems to me to imply an acceptance that the experiences are, in fact, paranormal," Shushan said. "So I would replace it with 'are reported.' There are essentially two possibilities as to why 'paranormal' occurrences are reported cross-culturally: (1) the human brain universally creates similar types of hallucinatory experiences in response to similar kinds of physical and/or psychological and emotional stimuli; or (2) they're genuine experiences of what they purport to be. In either case, whatever the base or 'core' experience is, it is invariably colored and even co-created by cultural and individual particularities. Scholars and scientists often try to reduce such phenomena down to a single prime-mover factor: put crudely, they essentially argue either that they're all in the brain (neurophysiological), all in the mind (psychological), or genuinely paranormal. One of my main arguments – going back to my work on ancient Greek and Egyptian dream diaries – is that unusual experiential phenomena have three layers of meaning and interpretation: universal, cultural, and individual."

Strange phenomena that occur cross-culturally also tie into folklore motifs. The "old hag experience," which is where an entity paralyzes them when they are asleep or awake, appears as the Mara of Sweden, bei Guai chaak (sitting ghost) in China, kanishibari in Japan and other names and cultures. Professor and author David Hufford covers this in his book *The Terror That Comes In The Night*. Hufford argues that certain parts of these supernatural occurrences are culturally derived, but the experience itself is not, and is independent of culture. Hufford states these phenomena include complex and stable patterns found in a variety of cultural settings, independent of explicit cultural models, experience itself plays a significant role in development of supernatural traditions, and the frequency the experience occurs is shockingly high; "the old hag," for example, 15 percent of the general population has experienced a like phenomenon. Memorates are an oral narrative from a memory relating to personal experience or legend. Hufford defined the term as "a story told as a personal experience believed to be true." Hufford considers these phenomena to be core-spiritual experiences, which includes a core foundation from which some supernatural beliefs can be developed by inference.

"David Hufford's work on sleep paralysis is a great illustration of this," Shushan said. "He identified the phenomena as a genuine cross-cultural experience, marked by a combination of sub-experiences such as an entity sitting on a person's chest, suffocating them or holding them down; inability to move, speak, or cry for help; fear and panic; and sometimes sounds such as voices or roaring. Hufford found that while these basic elements are stable across cultures, the identification of the entity changes according to local beliefs. So in Newfoundland it's a witch they call the Old Hag, across Asia it's identified as different types of 'sitting ghost,' and in the modern Western world it seems to account for at least some reports of extraterrestrials. Hufford's research was key in lending support to what he calls the experiential source hypothesis: that paranormal, spiritual, or religious beliefs are often rooted in extraordinary experiences rather than the other way around (i.e., that the experiences arise from prior belief). And this is regardless of whether or not the experience is genuinely paranormal."

Aquaman is a half-metahuman born of royal Atlantean blood, and half surface-dwelling human. He holds the abilities and characteristics of what could be described as the Greek God of Poseidon. Poseidon was the God of the sea, protector of all waters, and sailors relied on him for safe passage. Like Arthur Curry (Aquaman), he fought for possession of a city – in his case Athens, and Aquaman fought for the throne of Atlantis. Poseidon was also moody and temperamental like Aquaman, and they both carried a trident as a weapon.

Now, while we don't know of any "real" sea-dwellers that possess these powers, we do know that there are creatures of the deep that hold "superpowers" and abilities that humans do not. The octopus is an example, as they are able to adapt to their environment by changing color and camouflage. The *Turritopsis* jellyfish (*Turritopsis dohrnii*) can revert back to a "polyp" stage when injured, and restart life, which is basically a form of immortality. Starfish can regenerate limbs. These are just some examples of life on Earth that hold "superpowers" that sound like science fiction when compared to what humans can do.

If animals and various other life forms already on planet Earth can have special abilities, why can't humans? Is there possibly an inference to this story to potentially years of cover-ups that may have stagnated our growth as humanity?

"We have projected so many anthropocentric ideas onto it that we may find ourselves confused and correspondingly angry when we at last begin to see its general outlines," Levenda said. "Jesus may once again find himself being arrested, tried, and executed (especially in the American judicial system). Intellectually, we know that the Earth revolves around the Sun, and that the Sun is the center of our solar system, and that our solar system is just one of many millions of solar systems in a corner of the Milky Way galaxy, which is itself lurking in a corner of the known universe in the company of billions of other systems, but as much as we know all that, we still operate as if we are at the center of the cosmos. When we do that, we make certain assumptions about what is real and what is not; what is normal, and what is not; what is acceptable and what is not. We measure everything against a very meager metric of the state of human knowledge and awareness in the 21st century. It's not so much a cover-up, a deliberate shielding of the truth from the public, as it is a failure of imagination, an inability to reconcile what we experience with the intellectual systems we have developed for interpreting reality. Technologically, we are growing by leaps and bounds. There is no stagnation when it comes to AI, robotics, space travel, etc., but there is stagnation spiritually: we are still the cavemen we were when we encountered the Neanderthals, the Denisovans, etc. We still use violence to solve problems. We fetishize weapons systems from handguns to cruise missiles to nuclear bombs. That was the message brought to us by the alien portrayed by Michael Rennie in *The Day The Earth Stood Still* all those many decades ago, and we still have not embraced that message. What if Jesus had had a Klaatu all those centuries ago?"

The Nordic God Thor's origins began in the Roman era as he first was considered a deity in the process of *interpretatio graeca*. This method was used to interpret, or better understand, the mythology and religion of other cultures by using a comparative methodology to ancient Greek religious concepts and practices, deities, and myths/equivalencies. For example, you could compare the Nordic god Odin as "Mercury," Thor as "Hercules," and the god Týr as "Mars." There are various similarities and parallels in Egyptian, Greek, Nordic and other mythologies and Gods; think of it as comparable to the multiverse idea we see played out in comic culture. Just like you can have an Earth 1 Spider-Man and an alternative universe version of the same person, you have various Gods appearing in various cultures' mythology. Or what we have here is the *same Gods* being interpreted differently in cross-cultural phenomena.

The Thor we've come to know (the god of Thunder) is from the Viking age of mythology. Stone images and runestone invocations depicting Thor and his legacy exist in Denmark and Sweden. Is the lasting legacy of Thor just a legend, or are these stone tributes an ode to the son of Odin existing beyond just Scandinavian folklore? If Thor was a God or someone who appeared in the flesh and blood, could it be said he *lives* on today? If these various Gods from multiple cultures were once worshipped and lauded, but don't "exist" in the flesh today, does that mean they are gone? The legend of Thor lives on today through popular culture via Marvel Comics and the popular movies associated with the character. Is this modernized version of a God "living" in current times? Is pop culture's portrayal of iconic characters like Thor, Hercules, and others simply a new form of worship? Comics and cinema are just new mediums of storytelling, and today we are still telling these stories that began as legends. Legends never die, they just reappear in new forms of discourse.

In abduction cases, there have been reports of a race of tall, pale humanlike figures with blue eyes and long blond hair that are called "the Nordics." The Nordics sometimes appear in the UFO alongside traditional greys and other extraterrestrial beings, and other times the Nordics are the lone race interacting with the abductee. The Nordics get that moniker from their Scandinavian physical characteristics and "magical" aurora. Could the idea of the Nordics be an example of Gods from mythology existing? Or are the Nordics another offshoot of the idea of elves/fairies with their "magical" allure and physical descriptions? The idea of a grey alien, a reptilian and a Nordic-like God all being part of the same abduction experience may sound far-fetched to some, but like all abduction experiences in general, we have the same phenomena happening cross-culturally throughout different time periods.

In various religious texts, you have "angelic" visitations, and in European folklore there are tales of the fae, which is a supernatural "fairy" (nymphs, gnomes, and dwarves are also considered to be of a fae race), and the lore also includes abductions, lost time, and guarded, hidden (underground) realms. Legends also said the fae are the children of the goddess Danu. Danu is the goddess of the Earth, fertility, and wisdom of the Celtic people. Danu, like the Greek God Pan (more on him later), deals with ruling the wilderness and Earth, and both have fae-like guardians protecting their realms.

Folklorist and PhD Peter Rojcewicz has researched and taught international fairy tales and cross-cultural manifestations of the mythic imagination. He was trained in folklore, English literature, Jungian depth psychology, and Eastern philosophy and religion. Rojcewicz sees a connection between the alien abduction phenomena and fairy lore of the past. His studies suggest that we are looking at a current continuation of a much older narrative, and people have been having these like-minded experiences for the longest time, but just describe them differently. He also advocated for the Extraordinary Encounter Continuum Hypothesis (EECH), which refers to human confrontation with the anomalous, which includes beings (aliens, monsters, fairies), entities (energy forms, ghosts), objects (UFOs), or unusual lights (orbs). The EECH accounts for the cross-cultural distribution of these extraordinary beliefs, allows comparison to a diverse order of belief systems and generalizations, predicts and explains the nature of the unorthodox belief in logically consistent form of scholarly inquiry, and produces an operative definition that more accurately reflects the nature of non-ordinary experiences.

Did the believers of the fae experience a fairy-like phenomenon, or was it an alien abduction that they misidentified with their own cultural beliefs?

A cultural source hypothesis states that traditions emerge and are dependent on the culture that produces them. When applying the cultural source hypothesis to paranormal experiences, the explanation of the unexplained is culturally derived. For example, a luminous being from the sky may be viewed as an angel from a Christian standpoint. An experiential source hypothesis looks at the unexplained with minor cultural awareness, and beliefs are bound to not being able to explain the paranormal experience, so rationally it is "unexplained."

What if superior beings had nefarious means? Psychic-based abilities like remote viewing and telekinesis have been studied by the CIA, and people with these powers exist. What if they were to use these abilities to do harm? What if terrorists were able to harness meta-humanlike powers? We see this idea play out in popular culture through the idea of super villains, and the storylines that show when the good guys are manipulated to do bad things. This also goes with the fear of keeping God happy. Do you want a vengeful God that spreads disease and destruction because his followers didn't follow his teachings, or do you want a happy God that will spare the masses?

The idea of heroes becoming villains; what ramifications do "Gods" being bad have on the rest of us? In DC comics, there is a storyline that depicts evil versions of heroes that begins with evil the Batman Who Laughs (created by Scott Snyder) who comes from the multiverse and is a horrific hybrid of Batman and the Joker. In the *Batman/Superman* comic by Josh Williamson and drawn by David Marquez, the Batman Who Laughs is spreading his evil and forming alternative evil versions of Superman, Shazam and Wonder Woman with the ultimate goal of creating an evil version for each hero. The idea of any army of evil "Gods" is truly terrifying. In past mythology and religions, there have been tales of vengeful Gods and deities. This also plays into the idea that the all-powerful could be corrupted.

"So many stories from antiquity resonate today because they deal with timeless and innately human themes," Marquez said. "One of the oldest is the idea that absolute power corrupts absolutely. When given power over others, it's an inherently human urge to abuse that power. There's a very good reason why benevolent rulers are so revered in our stories – we recognize the sheer force of character and will required to resist that temptation. And superhero stories are our modern mythology – we use these immensely powerful characters to examine our own fantasies and our own flaws. If we had the power to move planets, how many of us could resist the temptation to use that power selfishly or destructively?"

One of the mass appeals of superhero lore is how they can connect these superhuman aspects to our basic humanity. We as consumers speak when Marvel movies and franchises are some of the most successful in Hollywood history. There are reasons people can relate to an Iron Man, even though most of us will never have his wealth or genius-level IQ. Fans can relate to the Incredible Hulk even though we don't have the ability to transform into this hulking monster full of super strength and basically impervious to pain, but we rally around the idea of an average person being able to transform into something greater.

"History and mythology are full of examples of the superhuman, and even today we hear stories of people capable of incredible feats: the mother who lifts a car to save her baby; the child prodigy; the miracle healer. In many cases there isn't any way of really knowing how true or exaggerated these claims may be, but for most of us they seem just plausible enough to be believable," Marquez said. "But more important to me than the historical or factual accuracy of these stories is what these stories tell us about ourselves, about what it means to be human. The mother lifting a car, something no human should be capable of, to save her child reflects the desire to protect the innocent. The child prodigy reminds all of us of the difficulty of learning, our pride in accomplishment, and our jealousy of those who seem to excel without any effort. And the miracle healer reveals our fear of death, our desperation to escape the inevitable when no hope remains, our longing for salvation and deliverance, and our hope that there is in fact more to this world than what we can see, touch, and hear."

These attributes and origins of God-like characters have been passed throughout time through mythology to modern-day pop culture and comic book superheroes. The basis of a lot of these stories remains intact, but throughout time certain aspects have changed to reflect modern society, which begs the question if our religion and history have gone through the same revisions?

The theory of gradualism is defined as the slow and gradual changes that happen within an organism or society to make a better environmental fit for animals and humans, such as a tiger developing stripes over time so they are better able to hide in tall grass. Think of human evolution evolving to better suit the inhabitants' adaptability to survive. If animals have adapted and evolved, who is to say we as humans have not, or have we not been able to tap into these abilities?

The X-Men are a superhero group created during the Civil Rights movement as a brilliant parallel to the racial inequality and persecution that prevailed in that time period.

"I couldn't have everybody bitten by a radioactive spider or zapped with gamma rays, and it occurred to me that if I just said that they were mutants, it would make it easy," Stan Lee, creator of X-men/Marvel Comics said in a 2000 interview with *The Guardian*. "Then it occurred to me that instead of them just being heroes that everybody admired, what if I made other people fear and suspect and actually hate them because they were different? I loved that idea; it not only made them different, but it was a good metaphor for what was happening with the civil rights movement in the country at that time."

The mutants were a race of people cast out and feared by the mainstream because of the way they were born or look or their culture. There are two sides here, and both are facing forms of oppression. One side would like to show the rest of mankind that they should not be hated and should be left alone to live in peace. The opposition wanted to rise and defend by fighting oppression and marginalization by any means necessary.

"The X-Men are hated, feared, and despised collectively by humanity for no other reason than that they are mutants, so what we have ... intended or not, is a book that is about racism, bigotry, and prejudice," X-Men writer Chris Claremont said in 1982.

The X-Men's teacher and benefactor, Dr. Charles Xavier, adopted the nonviolent philosophy incorporated by Dr. Martin Luther King Jr. as well as the self-improvement mantra of Booker T. Washington. On the other hand, Magneto attempts to change "by any means necessary," a line that Malcolm X was notorious for saying; thus, the distinction associates Magneto with the "radical" activist of the Civil Rights era.

The Nation of Islam used "X" to replace the American names with the ones they were born with during this time period. Most X-Men discuss the moral dilemma of how oppressed people should go about obtaining equal rights, through gradualism or radicalism, as seen in Bryan Singer's *X-Men* film.

Malcolm X once said, "I am for violence if nonviolence means we continue postponing a solution to the American black man's problem just to avoid violence."

Magneto refuses to wait for social change to happen on its own and wants to begin to implement it immediately by attempting to change all of the world leaders into mutants themselves so that they can begin to make laws that benefit both.

"Well, I think beneath the costumes and the spectacle and the fighting and the fun, there's an underlying philosophy about prejudice, about feeling outcast, fear of the unknown, trying to find your place in the world – very universal concepts that people have been attracted to," Singer said in a 2000 interview discussing the film as an allegory for a bigger dilemma.

The moral dilemma of how oppressed people should go about obtaining equal rights – through gradualism or radicalism – is at the forefront of these stories. While Xavier and Magneto use different means for accomplishing the same goal, Singer's film places emphasis on choosing a side – gradualism or radicalism. Adapt or die.

In the 2000 *X-Men* film Storm confronts Wolverine, and Wolverine responds, "You know there's a war coming. Are you sure you're on the right side?"

Storm replies, "At least I've chosen a side."

Singer wants the audience to think about the benefits and consequences of both radicalism and gradualism.

This makes a basic argument about cause of effect: *The unknown leads to fear; fear leads to hate; hate leads to violence.* Look at these quotes from various X-Men movies:

"You see, I think what you really fear is me. Me and

my kind. The Brotherhood of Mutants. Oh, it's not so surprising really. Mankind has always feared what it doesn't understand." (X-Men)

"Tomorrow, mankind will know mutants exist. They will fear us, and that fear will turn to hatred." (X-Men: First Class)

"Since the discovery of their existence they have been regarded with fear, suspicion, often hatred." (X-2)

Are there ways to turn on our own "hidden" superpowers? Is there a way to access these alternate channels without the use of DMT? Can we enlighten ourselves to open our own consciousness to potentially initiate "contact"? Our current state (human form) may not be equipped to handle the travel to other universes, but we may still have the potential to unlock those possibilities.

"My semi-informed opinion is that the field that we access via deep meditation (and some faster techniques) utilizes some of the same infrastructure (frequency, substrate) that 'they' do (mentally), and the craft does (by serendipitous manipulation of the local field), but no matter how informed, it's still just an opinion," said ex-Navy/*Nimitz* witness Sean Cahill, who also is a meditation specialist. "Resonance, frequency, harmony and vibration; this is how our reality operates ... When I strip myself of belief and bias, I feel like everyone everywhere is trying to interpret the same signal as experienced through the lens of their memory, culture, and fear. This means their location, family, language, religion or the science taught to them molds their world view as does yours and mine. If by our current state you mean this teetering between hope and fear, I am with you. The way we think and feel about each other needs to change. It needs to open. Knowing when to take a punch to forward the conversation is far more courageous than throwing one first."

If the phrase "life imitates art" is symbiotically true, would you also wonder if the relationship between the two are interchangeable? Does art imitate life too? Absolutely. Origin stories and influence from "true-life events" have been plucked from myths, folklore, history, and religious texts and been used to base fictional stories around.

When a movie is adapted from a book or comic, a lot of the time it is based on the source material. The source material being the format that the art originated in. Think about when a moviegoer sees Marvel's *The Avengers*. A lot of the film's plot points come from the source material, which in this case is various comic story-lines and, in some cases, stricter, single-story adaptations.

The origin myth of the Tower Of Babel from the book of Genesis, which explains why the cultures of the world speak different languages, repeats in comparative myths in Mexican culture, Nepal, Sumerian stories, and other cultures. In the Tower Of Babel a human race in the generations following the Great Flood (another origin story that launches the "universes" of basically every culture and civilization starting with the first Sumerian texts) travels to the land of Shinar, where they agree to build a city and a tower tall enough to reach heaven. God had other plans. God confounds their speech and spreads them around the world. The Tower Of Babel has been associated by some modern scholars with known structures like the Etemenanki (a ziggurat dedicated to the Mesopotamian god Marduk in Babylon).

In the DC Comics Justice League storyline Tower Of Babel, written by Mark Waid, the story takes the idea of de-powering humanity like a "God" did in the original story, and leaves the heroes floundering. Ra's Al Ghul hacks into Batman's hidden records, which conceal the weakness of every member of the League that Batman can use if one of the members ever goes rogue, and acts out the plan to disable the Justice League while he tries to reach his goal of reducing the world's population. Ra's Al Ghul follows this out along with his attack against the language centers of all humanity, using a specially designed tower to generate a low-level sonic signal that causes written language to be scrambled into nonsense, which sends the world's communication systems into disarray while he tries to complete his global goal. Sound familiar? This story expertly shows Batman's expert contingency planning, deep paranoia, and wariness of meta humans.

Martian Manhunter (J'onn J'onzz) is one of the most powerful beings in the DC Universe and one of the original members of the Justice League. His abilities include telepathic communication, the ability to summon others/implant images and ideas in the minds of others (some were apocalyptic warnings of the end of the world), shapeshift, and telekinesis. Now, take the abilities of this fictional character and compare them to stories of alien abductions and the perceived abilities of extraterrestrials, and ask yourself if Martian Manhunter is possibly based on these same stories?

The iconic Wonder Woman is based heavily off Ancient Greek mythos, and Greek mythology is ingrained in Wonder Woman's fictional world. Diana Prince's (the name she adopts once becoming a part of society) origin story relates that she was sculpted from clay by her mother, Queen Hippolyta, and was given a life to live as an Amazon, and was given superhuman powers by the Greek gods. Another version of her origin story dictates that she is the daughter of Zeus and Hippolyta and is jointly raised by her mother and her aunts on Themyscira. Themyscira is the Amazonian island, which is another world that exists on Earth and is hidden/protected by the Gods. The idea of a fictional civilization in Themyscira isn't an alien idea when presented with the real-life mysteries of the Greeks, ancient/hidden civilizations of Lemuria and Atlantis, Hollow Earth, and the general theory that other worlds exist among us, but are blind to the unseen eye.

Marvel's Dr. Strange is a character that intersects the metaphysical world and the physical. Dr. Strange "flips" from being an egotistical surgeon who loses the ability to operate after a car crash severely damages his hands, to an enlightened master of magic to defend the world from mystical threats. After his accident, he searches the globe for healing and encounters the Ancient One (the Sorcerer Supreme). Strange becomes a master of the mystical arts, acquires an assortment of mystical objects, including the powerful the Eye of Agamotto and Cloak of Levitation, and lives in a New York City mansion called the Sanctum Sanctorum. In some of the comic stories Dr. Strange encounters Egyptian myths, Sumerian gods, and Jungian archetypes.

In *The Flip* by Professor Jeffrey Kripal, he references the idea of a reversal of perspective or a flip. In the book Kripal details academics, scientists and historical figures who have experienced things that would flip their world view. In Dr. Strange's case, he goes from a man of hard science and skepticism, but then is "flipped" to accept a new reality of the mystic arts and higher consciousness. The 2016 Marvel Studios *Dr. Strange* movie showcases this flip nicely and further demonstrates that bridging the gap between science and paranormal isn't as hard as humanity makes it out to be.

So, to sum it up, we have (just to spotlight a few):

-God-like figures based on mythology like Wonder Woman, Aquaman (representative of marine species having unique abilities we don't) and Thor.

-Biological entities like Swamp Thing and Poison Ivy, who can manipulate part of the Earth (think the green man or the God Pan).

-Heroes made from science like the Hulk and Captain America.

-Alien saviors like Superman or Martian Manhunter.

-Batman representing the idea of God vs. man.

-Mutated genes that make once human beings into mutants with special abilities who get ostracized and discriminated against in the X-Men.

-A half-robot/half-man-like Cyborg, which could be

used as a comparison to robotics, advanced computeriza-
tion, and robotic limbs being used in the medical field.

-Humans using science and innovation to appear
"super" in nature like Iron Man.

-Accidental abilities from radiation in Spider-Man.

All of these fictional elements *potentially* exist in our
reality.

> *"Mythology exists at a level our social reality over which normal political and intellectual action has no power." – Dr. Jacques Vallee*

IF THINGS DON'T EXIST in the organic, biological bodies that we are accustomed to, or don't always appear visible to the eye, then do they exist?

We see a movie, and we know the characters are fictional, but they exist to us since we see them as a physical manifestation. UFOs appear to be a physical object you can touch, but since there hasn't been an official disclosure to what they are, do they exist beyond our consciousness?

The UFO community for the longest time has been relegated to the shadows and considered by many to be subcultures full of pseudoscience, New Age nonsense, and overall weirdness. Since many in the UFO community have come up through this subculture, there has been a shift in the way the subject has been perceived in tone, and media attention.

"Our 'spectrality' goes beyond simple cultural acceptance," said Terra Obscura blogger and VICE contributor MJ Banias. "I think that is an important battle, but the war is much more complicated. We are a community of people that live in a very curious dualism. On the one hand, we accept that UFOs may be some 'intelligent non-human other' and that our reality is much stranger and more negotiable. We accept that something much grander than us exists and it interacts with us regularly. At the same time, we continue to exist within our 'normal' daily reality. We go to work to pay our bills, drive our kids to karate class, and still function in arbitrary social, economic and political frameworks. We are ghosts because we can't seem to choose a side. We live in this gap between two very different worlds, a sort of liminal ghostly realm, where our identity exists and does not exist simultaneously."

Where do we draw the line between UFOs, paranormal and cryptozoology? Or do they all intersect?

"It's just the nature of the unexplained in general," documentary film director Seth Breedlove said. "It's funny, I was just speaking at a MUFON conference and

I mentioned that I don't draw a dividing line between cryptid or ghosts or UFOs. I'm fascinated by all of it, and there are so many common factors running between all these different phenomena that I just say I'm a fan of the unexplained."

Paranormal is roughly defined as denoting events or phenomena that are beyond the scope of scientific understanding.

"I don't like the word paranormal because it suggests that something that we see in the universe, and it exists in the universe, it's not supposed to be in the universe," said scientist for US Army Space and Missile Defense Command and Skinwalker Ranch lead investigator for the History Channel Dr. Travis Taylor. "What I saw (at Skinwalker Ranch) was within our universe, and it was 'normal'; it was just something we don't understand and don't know what it is. Absolutely, without a doubt we have scientific instruments that detect and measure, multiple witnesses and cameras, and multiple occasions of phenomena that cannot be explained by human technology. This doesn't mean it can't be explained through a future or better understanding of physics."

Veteran investigative journalist Leslie Kean's groundbreaking book *UFOs: Generals, Pilots, and Government Officials Go on the Record*, followed by her co-authoring two major *New York Times* news stories about the Pentagon UFO program and Navy pilot sightings, has helped provide legitimacy to the unexplained. Her latest foray into the unknown, the 2017 book *Surviving Death: A Journalist Investigates Evidence For An Afterlife*, continues her search for answers within these extraordinary topics. Kean is working on an upcoming documentary series based on *Surviving Death*, which includes coverage of past lives among other research areas. Both her nonfiction bodies of work are important in that they tie together multiple phenomena. Her research has shifted from UFOs to consciousness/afterlife, but did it feel like a logical progression?

"It did feel like a logical progression, partly because the topic of survival past death is something I've been interested in for a long time, throughout the years I was reporting on UFOs. It was sort of in the background for me, and at one point I assisted producers making a documentary on the topic, so I was exposed to a lot of the information and experts in the field. In doing that," said Kean, "it wasn't so much that I saw a link between these topics and UFOs, but I was very fascinated by the question of possible survival after death, and I knew there was a lot of research I could do on it. The difference was I didn't know much about the topic yet. My UFO book represented the culmination of ten years of work, but this was different. I was jumping into something that was relatively new to me, and it was much more of an exploratory journey that took place while I was researching and writing *Surviving Death*. When I started, I didn't know how it would come out, and never expected to have the experiences I had along the way."

Some of the things she has witnessed makes UFO sightings seem "normal" by comparison. Kean has witnessed a physical medium elicit ectoplasm, which would make for good evidence, but begs the question: would it ever be possible for a medium to do it on camera, or live on camera, to help convince the majority that this is legit?

"All of us would love to have that, and theoretically, yes, it is possible. Historically, legitimate photographs were taken of ectoplasm, but others were faked, which creates major problems. To do this now, first of all it's a matter of finding a medium who is trustworthy," Kean said. "We would have to know there is no fraud. The medium would have to be developed enough to handle low-level light in the room and the equipment needed to film, without impacting the phenomenon itself. Ectoplasm is very sensitive to light. Then, this medium has to be willing to allow this. There are many reasons why a legitimate physical medium would not want to undergo the kind of controversy that would come from engaging in something like this."

. . .

Kean has been working with Stewart Alexander, a British physical medium who has been practicing his mediumship for forty years. She has seen with her own eyes the manifestation of a "living" human hand form out of ectoplasm on multiple occasions, as have many others, and has touched the hand. These events occurred under controlled conditions so that Kean could be sure of their authenticity. Regarding filming this, Alexander is not in favor of the idea due to its potential professional and personal ramifications. Kean explained that Alexander is a private person and doesn't want to draw disruptive attention to his small, regular weekly circle and his family. The media hoopla would be destructive, and the skeptics and scientific community would no doubt claim the whole thing was bogus, as they have done to the detriment of genuine mediums in the past. Alexander wouldn't want this hanging over his legacy, leaving his heirs to deal with such questions. The stress of this could affect the sensitivity he needs to maintain in order to continue with his mediumship.

Orbs and voices of deceased relatives are sometimes associated with other areas of paranormal. Near-death experiences and other sensory phenomena all tie in to these threads. Are these orbs "souls" that live on in the universe?

"I know that people around the world and throughout history have reported NDEs, and my research has convinced me that NDEs frequently contribute to beliefs about an afterlife across cultures," said author of *Near-Death Experience in Indigenous Religions* Gregory Shushan. "I also know that such reports often describe encounters with deceased relatives, for example – but I have no way of knowing whether they really are the surviving consciousnesses of deceased human beings, or any objective entity at all, or indeed whether they are hallucinatory. The same can be said of any NDE element. Even if I were to experience them myself, I would probably still question them (as I have with apparently genuine extraordinary experiences with psychedelic drugs, meditation, and mediumship experiments). Same if I were to encounter orbs speaking in the voices of my deceased relatives. Although, if I were awake and in my right mind, admittedly that would be pretty compelling! Especially if someone else was there to independently corroborate. In any case, I'm comfortable sitting with that uncertainty, in knowing what I don't know, and reserving judgment until I eventually find out for myself ... In both of my books, I talk about how the diversity of NDE accounts impacts materialist theories, but also how they impact metaphysical theories. Specifically, what kind of afterlife could be philosophically imaginable given both the cross-cultural similarity and diversity of NDE accounts? In the first the questions were in relation to ancient civilizations, and in the second to indigenous societies."

Ron Morehead has been in pursuit of Bigfoot for over 45 years at this point, and the veteran cryptozoology adventurer has evolved his focus on the elusive creature. His book *The Quantum Bigfoot* is a different take into trying to figure out exactly what Bigfoot is, and how spirituality can combine with newer science. For someone who has spent so much time in the woods all over the world researching Sasquatch, he's able to simplify the complicated concepts of quantum physics in the book.

"Because I'm not a physicist, I worked to keep the book in layman's terms," Morehead said. "About 10 years ago I began to try and understand how these beings are able to do what they do. Al Berry advised me years ago to always stay within science to stabilize credibility with academia. This brought me to quantum science, which is actually fairly new. Quantum physics is the same as spirituality, in my opinion. This took me back to my church days and many of my childhood scriptural memories. As I researched ancient texts more thoroughly, I realized the 'science' behind what we call miracles is actually quantum physics."

Morehead is best known for the famous Bigfoot recordings called the "Sierra Sounds," which feature the best vocal evidence to date that has baffled crypto-linguists. He has seen the animal once with his own eyes, and his beliefs on whether it is a paranormal or biological being land in the middle.

"They talk, leave footprints, defecate, and procreate ... that means they are physical," Morehead said. "However, I believe they are a remnant of foreign intervention into the genome of a primate, and the DNA was altered by an advanced technology. Over eons, some have been diluted down by cross-breeding with humans, thus making some more humanlike than others. This alien component gave them the ability to do things that we are currently trying to understand. I personally believe that quantum physics is the answer to understanding. Nothing is actually 'supernatural,' just our lack of understanding what is 'natural' for humans as a species. We just haven't evolved enough yet."

The idea of quantum computing, energy and studies are slowly becoming a reality, as science has opened up to the idea of studying quantum mechanics, which is defined as the body of scientific laws that describe the different behavior of photons, electrons and other particles that make up the universe. A quantum is the minimum amount of any physical entity involved in an interaction. Quantum mechanics, including quantum field theory, is a fundamental theory in physics that describes nature at the smallest – including atomic and subatomic levels.

He was young, brash and he thought outside of the box. He was called a genius, and a king of enterprise before most were even out of college. This young hot-shot thought outside the box, and made waves. The swag, and the goal wasn't self-serving though because this person truly wanted to help people, and make the world a better place. The technological innovations came, as computing, aviation, defense and cybernetics all had breakthroughs with his vision. He was a man, but when he strapped on the suit he became something else entirely.

– Tony Stark (Iron Man)

The University of Toronto once asked the question if tech prodigy Deep Prasad (then an undergrad) could "be the next Einstein."

At age 23, Prasad is the CEO of ReactiveQ, which aims to create the world's first quantum computer, as well as working on producing superconductors and meta-materials. The young star of innovation is also an avid UFO advocate; he's one of us. Part of ReactiveQ's mission statement states they are "actively engaged with engineers from TESLA, Lockheed Martin, Volkswagen and NASA in order to validate solution."

Being a head of a company complete with executives, shareholders, and funding from the likes of Bloomberg's Venture Capital arm, Prasad is unafraid to speak his views on a once taboo topic.

"I believe most of them are (OK with his beliefs), and I've never gotten 'flack' for my beliefs, not once, which is encouraging for someone whose career relies on having a reputation of scientific and intellectual competence alone in order to succeed with the many stakeholders involved," Prasad said. "My core team is super relaxed and open about the concept and have changed their minds a lot since I started talking about it, but they remain healthily skeptic until they are provided the hard material science data and other physical evidence the community requires to 'believe.'"

ReactiveQ is currently working with engineers from Lockheed Martin, TESLA and NASA. These specific companies could offer a lot in terms of ideas of reverse engineering or better understanding new technology, which begs the question was this a mutually beneficial undertaking that possibly can tie in with advanced "foreign" technology?

"I can't comment on the advanced 'foreign' technology aspect, but what I can say is that everything my company does in the eyes of the public in the near future will be 100 percent strictly related to terrestrial non-'foreign' projects," Prasad said. "The specific reason was that my goal before I really got obsessed with this UFO subject was to accelerate humanity's technological development by a factor of hundreds of billions of human hours. That is, I want to provide simulation capabilities so good that one can eventually algorithmically discover new engineering breakthroughs within minutes without physically having to go through the years-long trial and error process of building out different engineering prototypes. Now that I know what's possible, I can actually use humanity's ability (how far along they are in their ability to replicate 'foreign technology') as a measure for how successful or useful our company is at any given time."

One of the targets of ReactiveQ is the creation and exploration of meta-materials that Prasad states will require breakthroughs in quantum physics, material science, quantum computing, advanced manufacturing technology and capital-intensive resources. If humans tried to re-create a craft like the Tic-Tac UFO at *Nimitz*, it would take collaboration from industry leaders. The ideas of consciousness and meta-materials are more accepted in UFO studies, but these ideas still present these challenges to the scientific community.

"I think we have to first define the term consciousness more rigorously," said Prasad, who is also an experiencer. "The good news here is that should we encounter an advanced non-terrestrial species, they would, according to us humans and our simple ways of classifying intelligent life, fall into two categories in our eyes: biological or nonbiological. Now, when or if society accepts that these entities may actually contact humans from time to time, then the next obvious question is going to be how they get into our heads? The technology they're using to communicate with citizens of Earth seems to have no bearing or care for whether the person is awake or not. This should throw our entire understanding of reality into question for anyone paying attention, as what you dream about should only be accessible to you."

Hopefully thoughts like this will get the community thinking about the importance of consciousness in their own ways. The idea of mankind rallying together behind a common cause has been seen throughout history. Sometimes it is just an average joe trying to make a difference through activism, or larger campaigns of change.

In the 2017 *Justice League* film, Bruce Wayne is met with a new world (threats) that is beyond what he believed was the realm of possibilities. He realized he couldn't save the world alone, so he sought extraordinary individuals to form a team to usher in this new era.

He came from a broken home. He was able to use his passion and energy to create resources that would later be helpful down the line. He traveled the world and learned his craft. He realized he couldn't fight this battle alone and needed the help of those with more to offer; with different skill sets, with different "powers" and who were willing to sacrifice. But would they risk it all to join his league?

— Bruce Wayne (Batman)

Tom DeLonge left the spotlight of superstardom when he walked away from his multi-platinum pop-punk act blink-182 to start the machine of To The Stars/ To The Stars Academy of Arts and Science. Somehow, he was able to gain access to some high-profile people in the Pentagon, who were able to listen to his vision and point him to the right people. What would eventually come to fruition was a concentrated effort in trying to change the world with To The Stars Academy of Arts and Science, which was part science/research, an entertainment division to bring DeLonge's fictional properties associated with the phenomenon aside, and an aerospace area to eventually try to innovate. The team consisted of Luis Elizondo, a former Pentagon official who headed the Department of Defense's Advanced Aerospace Threat Identification Program (AATIP), but grew weary of all the governmental red tape. There was also Jim Semivan, who left his job at the CIA after serving for over 20 years. Steve Justice joined after 39 years at Lockheed Martin (31 of those with Skunk Works). Dr. Hal Puthoff (CEO of Earth Tech International / director of Institute for Advanced Studies in Austin) is no stranger to the UFO world, as his history as an advisor to NASA, the DOD, and his research on "cutting-edge technologies" are all well-documented.

This was DeLonge's league of his own he assembled. Small in size, but large in scope, TTSA have slowly begun their journey, but DeLonge and company aren't the only ones trying to study and innovate metamaterials.

Businessman and entrepreneur Robert Bigelow is best known for starting Bigelow Aerospace in 1999, which has made huge strides in space exploration technology and has been contracted to do work for NASA. Bigelow has been a long-time believer that aliens visited earth. Bigelow Aerospace isn't the only company Bigelow is known for, as in 1995 he started National Institute for Discovery Science (NIDS). NIDS focused on scientific studies of fringe topics like UFOs, cattle mutilations, paranormal occurrences, and most notably Skinwalker Ranch, which featured all of the above anomalies. The high strangeness that occurred allegedly garnered the attention of the US government.

Hunt For The Skinwalker was a game-changing book by journalist George Knapp and NIDS investigator Colm Kelleher, PhD, that blew open the paranormal plethora of scientist-baffling anomalies that have haunted a ranch in the Uintah Basin.

The book does more than just detail the plethora of paranormal activity that went down at Skinwalker Ranch and the scientific team who investigated UFOs, light orbs, giant (bulletproof) animals, shadow beings, cattle mutilations, portals, Bigfoot and other anomalies all hosted at this remote part of Utah. Journalist George Knapp and NIDS investigator Colm Kelleher, PhD, who witnessed the experiences do a superb job of providing background of the area through history, Native American mythology and connecting it to other high-traffic paranormal hot spots around the world. In addition to connecting to other similar cases and folklore, the book runs down every possible hypothesis from outer worlds, dimensions, hoaxes, military experiments, extraterrestrial intervention, and more.

"In some ways, it's been harder for the public to learn about the activities on the ranch than about some of our government's most secret military bases," said filmmaker Jeremy Corbell, who made the documentary of the same name. "This is due to a number of reasons: the remote location, active security measures, tribal law, private ownership, liability concerns and even superstition. There were obstacles, yes, and there has been a veil of secrecy over the ranch for over two decades. From the Defense Intelligence Agency's study of the property through AAWSAP (Advanced Aerospace Weapon System Applications Program), to the NIDS study privately run by Robert Bigelow – to what's going on now under the new ownership – we have learned more about the nature of the phenomenon from these scientific endeavors than ever before in history."

There's another side to this story being exposed by UFO researcher Erica Lukes and former Bigelow Aerospace Advance Space Studies investigator Chris Marx.

Lukes has been conducting her own research of the ranch, which includes on-site studies, and has used Marx's renewed interest in the ranch to add another layer to this complicated story. Lukes recently released the first video interview with Marx, "Skinwalker Ranch:

A Real Perspective," which details just some of the phenomena Marx experienced from his six-year study on the ranch. Marx resurrected his passion for the ranch when he saw Corbell's *Hunt for the Skinwalker* documentary, which prompted Marx to "rectify the madness." Marx believes the film does a disservice to the ranch by pushing a threat narrative.

The big divide is on tone and perception. In Lukes' video, Marx describes the ranch as "sacred," a "very beautiful place," and "an environment like no other," which is a stark contrast to Corbell's film that paints the ranch as nightmarish in landscape.

"It's clear to me that they take a very different angle on the property," Lukes said on the *Hunt for the Skinwalker* film. "They clearly view that place as a 'threat,' and while that angle might line the pocketbooks of government contractors, I certainly feel it is a disservice to frighten local residents. To promote the idea that encounters cause debilitating effects on witnesses without doing a thorough investigation on other factors that might contribute to disease is unconscionable. I agree 100 percent with his (Marx) statements and applaud him for having enough respect for the ranch and neighboring community to come forward and tell the real story."

Marx does state in the video that being on the ranch

is "not without risks" and still suffers from some consequences like random crashes in his kitchen and misplaced items in his home. The video's big reveal is Marx explaining encounters with multiple entities, including his eyewitness account of a man shape-shifting into a wolf.

"I have a deep affection for the ranch and the surrounding area," said Lukes. "It's awe inspiring and perplexing. I have personally witnessed and documented things that are out of our realm of understanding, and I will always hold it very dear to my heart."

If we look at all the alleged occurrences that have been reported at Skinwalker Ranch, are any of them explainable from our scientific standards and everyday understanding of reality? The NIDS scientists couldn't control any of the variables on the ranch, but said they felt like the phenomena were controlling them. So any traditional scientific method (observe, question, hypothesis/testable explanation – a prediction based on that, testing the prediction, use the results to make a new hypothesis) were void since anything they tried to test couldn't be measured. If anything, the phenomena were messing with them, from equipment malfunctions to instances that defied possibilities.

Dr. Travis Taylor is no stranger to the world of the paranormal, as he has appeared on History Channel productions like *Ancient Aliens, The Tesla Files,* and *The UnXplained,* so Taylor was an obvious candidate to be the lead scientist on History's nonfiction series *The Secret Of Skinwalker Ranch.* Taylor was skeptical at first as to whether the phenomena were a natural phenomenon, a classified defense project or something supernatural in nature. Since taping, Taylor claims he now suffers "horrifying nightmares" and experienced radiation poisoning and sickness at one point.

Taylor's pedigree includes 25 years of work with programs for the Department of Defense and NASA, and he currently serves in the US Army Space and Missile Defense Command as the principal investigator for Quantum Entanglement and Space Technology. Taylor holds PhDs in optical science and engineering, and aerospace systems engineering; master's degrees in physics, mechanical and aerospace engineering, and astronomy; and a bachelor's in electrical engineering. His recent projects include advanced propulsion concepts, quantum teleportation, space telescopes, space-based beamed-energy systems, and "future combat technologies," which all sound like topics that sounded impossible years ago, or like science fiction.

"I'm also a science fiction writer; I've written like 23 best-selling science fiction novels," said Taylor. "So I have an interesting point of view, or more of an open mind/imaginative point of view of phenomena I see, and what I can tell you that the phenomena that was observed when I was there was beyond the physics of mankind."

There are some who believe the site has been used for weapons testing, and the anomalies on the ranch are manufactured by government programs. Taylor alleges this isn't the case.

"What I can say is I'm 80 percent certain it is not natural phenomena (and that's not say Mother Nature can't do amazing things), and 99 percent certain it is not man-made phenomena. As a person who does weapons testing as his day job, I'd tell you that it would be so highly crazy illegal that it is nonsense," Taylor said. "There would be people in jail. After what I observed my first day on the ranch, we had a long discussion that if what we were observing was man-made and someone was violating federal laws, we needed to alert the authorities if we could prove it was man-made, but I realized what we were measuring was even impossible for mankind to make, and at that point is when I dropped that line of discussion. I realized flat-out mankind was not doing this ... and it is probably a skeptic's coping mechanism, because I did it too; the first conclusion to an odd/strange thing is 'oh, that is a classified government program and they are doing human-testing on us.'

There were CIA programs in the 1960s/1970s that I don't think they are proud of where people were involved in those experiments, and so you look at it nowadays you realize we can't do that and won't get away with it. I'm thoroughly convinced this is NOT some top-secret weapons testing program. There's oversight committees in the Senate, and eventually somebody would say 'hey, wait a minute; y'all can't do that.'"

The ranch's new owner, Brandon Fugal, has taken a hands-on approach to investigating his property. Fugal, one of the most prominent businessmen and real estate developers in the Intermountain West, purchased the property from Bigelow in 2016 in order to investigate the strange occurrences of Skinwalker Ranch. Contrary to what some detractors have accused on social media, the involvement with the ranch isn't a financially motivated move, and he accepted History Channel's investigation on the grounds the show remained true to investigative integrity.

"During the entire four years of owning the ranch, I have yet to take one dime personally in connection with the property," said Fugal. "It is a money pit. Even if the show is wildly successful and continues into the future, I can't see any possibility of recouping even a fraction of what has been invested into the property in service to the research and investigation. My involvement with the ranch is a liability. In terms of financial motivations, my time is much better spent on my conventional business activities. The History Channel didn't reimburse me one penny regarding use of the vehicles and equipment and facilities there on the property. Although the team was compensated for their time, and there was a modest budget to assist with some of the experiments and testing, all of the infrastructure and primary equipment was bought and paid for by me over the course of the last several years. None of my people had any intention or aspirations to be on camera or be in a television show. In fact, it was quite the opposite. Most of them hated it, especially at first. This is entirely unscripted. One of my main conditions with a green to the investigative series was that nothing could be fabricated, exaggerated or contrived. It had to be 100 percent true. ... I am not trying to legitimize the ranch. I am simply allowing the public the opportunity to follow our research and investigation."

In the area, there are two indigenous tribes, the Utes and the Navajo, and there are stories invoking the feuds between the two, which include potential curses of a Skinwalker, as well as "sky-gods" and other Native American lore. Many of the sightings go hand in hand with Native American folklore, and some of the entities reported represented what could best be described as spirits or Native American ancestors. Folklore intersecting reality – fact meets "fiction."

But Skinwalker Ranch isn't the only area that holds high strangeness. There are multiple locations around North America and the world that have multiple reported sightings of UFO flaps, Bigfoot/cryptid sightings and general weirdness. The paranormal documentary *Hellier* started with exploring a rural area of Kentucky that reported sightings of "goblins" and eventually led to a case much greater that connected the surrounding areas of Ohio, Somerset, Kentucky and Point Pleasant, West Virginia (sight of the Mothman sightings and infamous bridge collapse). It wasn't just Hellier, Kentucky, that was prone to strangeness, but the surrounding Appalachian area and the massive underground cave systems that connect them all.

Season 2 of *Hellier* (2019) boldly ties all aspects of unexplained phenomena in a mind-blowing fashion that will make you question your core beliefs and the world we live in. What Greg and Dana Newkirk, director Karl Pfeiffer, Connor Randall, and Tyler Strand have presented here is a courageous journey that goes beyond just the case of "goblins" in Kentucky. If anything, *Hellier* weaves a narrative that expertly connects the dots between any and all high strangeness. Many believe they (UFOs, cryptids, ghosts, etc.) are all connected somehow, but *Hellier* was able to provide a greater map on how they are all connected. The inner skeptic in me at times felt me saying "no way" to myself at some of the theories proposed here, but when tying those notions with familiar literature, history, and knowledge, that "no way" turned into a "they may be onto something." They are onto something here; a weird way that UFO researchers, occultists, cryptozoologists, folklorists, historians, and academics can all come together and realize that, ultimately, they are all studying the same thing: a singular phenomenon.

Episode five connects the UFO to other phenomena in a way that will blow your mind if you are into mythology, UFO history and folklore. The reference to the book *Etidorhpa*, which was a book published in 1895 that tied into the cave mystery that Planet Weird was studying, stood out to me because it was a book referenced in Professor Jeffrey Kripal's brilliant book *Mutants and Mystics* and was suggested to me by him as well. "Etidorhpa" backwards is Aphrodite, and in the book the protagonist discovers a whole new world (and to the Earth's core) from the caves discovered in – Kentucky. *Etidorhpa* included ideas of practical alchemy, secret Masonic orders, the Hollow Earth theory, and the concept of transcending the physical realm, as well as meeting beings from the cave.

The town of Somerset is allegedly sitting atop all this quartz that some in the town believe is causing these vibrations that lead to high strangeness; makes you wonder what is under Skinwalker Ranch, and why it is advised not to "dig" there? Does this mean any paranormal sighting is basically a "glitch" in the universe, and these beings live among us in a different reality or it crosses universes? So, if these paranormal hot spots exist, there is something elemental underground that is tied to it. What would happen if you dig in Skinwalker Ranch? Digging is generally prohibited, and those who have tried have suffered odd and severe injuries. The shape of the basin where the ranch resides is interesting too.

"if you look at the Uintah Basin on Google Earth, to me it looks like an ancient meteor impact crater, and, in fact, it looks like it came from the east to the west at a low inclination," said Taylor. "And that's what splattered the salt flats to the west of the Uintah Basin, and there's Gilsonite all around the Uintah Basin, which typically is only found in a meteor impact crater; plus all of the petroleum that is underneath the Uintah Basin, there are a lot of geologists and natural physicists now beginning to think that impact craters cause a phenomenon that creates petroleum. If you look at this impact crater, the ranch is dead center (give or take), but it's pretty much dead center. And perhaps something to do with the bowl shape of the basin or whatever caused the basin made this the center or the nexus for whatever the activity might be."

Would hitting the wrong thing equal unleashing a paranormal Pandora's box of shadow beings, ghosts, orbs, mothmen, and other monsters? This scenario sounds like something out of science fiction, but what if certain elements do hold more power than what we believe? If drilling quartz that is buried underneath Somerset unearths some gateway to beings from another world, is this an experiment worth doing? What if the sightings of beings and cryptids are "just passing through" our world, and what would the ramifications be if we allowed hostile entities? The idea that they are "passing through" does match with reports at Skinwalker Ranch of giant wolves and ancient Indian chiefs being witnessed and disappearing into thin air, as similar sightings of Bigfoot all over the world have reported a similar phenomenon. This would explain why so many cryptids and unexplained creatures have remained so elusive. Maybe these trans-dimensional entities are supposed to remain secret from the human eye, or maybe some of those in power know what these locations hold, and are dedicated to keeping the status quo. Again, if you think hiding UFOs from the public is one thing, imagine having to explain monsters, ghosts, and other beings?

The name of the town Hellier is interesting. Hellier. The root being "hell." If this area is a "gateway" to another part of the world like many in the area believe, could this at one time have been considered "hell"? The "goblins" that guard the caves are referred to by some witnesses as "gatekeepers." So imagine, no prior knowledge of this, but if religious people saw these odd-looking things guarding a cave that went underground into the Earth, could a religious person see this as hell? Think of the idea of a Cargo Cult; a belief system among an undeveloped society that sees a phenomenon as of what they are familiar with. If an indigenous civilization has never seen an airplane, and the first time they see it they associate it with a bird, then to them that plane was a bird because they have no prior knowledge of aerodynamics.

Hellier continues to learn of allusions to a "green man" or a God of the earth in all their research. At one point they try to make contact with a dead "God" in Pan. In Greek mythology Pan is the god of the wild, shepherds and flocks, nature of mountain wilds, and companion of the nymphs. He's also described as looking like a faun (in Roman myths he is called Faunus). If "Pan" was this horned God of the Earth, does this mean if looking at it from a Cargo Cult perspective that the idea of the devil was misidentified from the beginning?

Is there a way to summon gods, aliens or even open gateways to other worlds? Experiencer Whitley Strieber's best-selling book *Communion* was a gateway to UFOs for many people, but Strieber himself may also be a gateway. If *Communion* was a call to arms for abduction experiencers and an introduction of phenomena to mass audiences, his book *A New World* is an awakening for those in the UFO community who refuse to continue to look at the universe from a different view.

In *A New World* Strieber makes the point that experiencers are left to fend for themselves, which means the idea of "disclosure" is an uphill battle. The secrecy is controlled by the visitors by integrating themselves into our lives more openly, rather than a full disclosure. In general humans are "passive to the phenomena," and the coming of visitors is when "evolution comes to a conscious mind." Another way we fight the idea of communion is "ego." Ego defends its own existence against the presence of aliens, which could explain how enlightenment and other ways are "contact." He also makes an interesting point that the Roswell crash was likely more of a "donation" rather than an accident; in turn it did what they wanted, which was to keep presence secret while trying to advance technologically.

Some of the better questions Strieber proposes are

about our perception of what we know as our world, and that there is more there than we know. Is this other world a companion universe? Is dark matter the mirror universe, and is this where visitors come from? Do humans with advanced skills come from it? There has been increased funding for studies regarding dark matter, and dark energy is not a taboo subject for the scientific community to research. For example, the University of Texas McDonald Observatory received a $40 million facelift to its telescope and will now aim to study dark energy. Dark energy, which is the enigma that is defined as an unknown form of energy that can supposedly permeate through all of space and, in turn, help expand the already expansive universe. Dark energy also assists in repelling gravity, which is said to take up 70 percent of the universe. The telescope at McDonald Observatory will have the ability to create a 3D map of the universe, view lights as old as 12 billion years old, see red-shift galaxies that can be up to 10-to-12 billion light years away, and have a range of fields that can observe 120 times more of the sky at night.

Strieber recounts that allegedly he has been able to, at times, see the other world exist while in this one. How he got there, or where this ability supposedly comes from is a mystery, but he did witness this "other world" that doesn't look too different than ours on an Indian reservation. Could this "other world" serve as heaven or possibly even purgatory? Is this other world a place where the dead, alien beings, and cryptids live? Witnesses at Skinwalker Ranch have sighted ancient Indian chiefs in the area that are as real as a biological lifeform, but were able to disappear into thin air. Coincidence?

One of his stories mentions when Strieber and his guests were visited one night, they all experienced knocks in three groups of three, which was nine distinct knocks on the roof, and almost sounded like three cries. There's a scene in *Hellier* while trying to channel an out-of-this-world being they heard the sound "bing, bing, bing" and advised it should be made in threes. Does all this make more sense now? Strieber references this as a similar ritual done by Masons for a 33rd degree Mason. Three-three-three done by three people would equal nine. Interviewing a Master Mason who would like to remain anonymous, he confirmed the rituals exist, and that the number three is a very astronomical and luminal number to Masonry. Numerology is a whole other story here, but you can make the argument that sounds and vibrations can elicit some sort of contact and possibly even open up doors to new worlds.

In a more disturbing part of the book, Strieber details human mutilations (homeless), which reflect the same phenomena as cattle and cat mutilations. The spine being removed is significant if you believe the spine holds an "energy" or a soul, and you could capture a soul (the *Tales from the Crypt* episode "Doctor Of Horror" is about this.) This all goes back to the theory of our soul is our consciousness. Now, some believe it could be extracted, and possibly extraterrestrials are doing this to animals and some humans. It hasn't been fully explored yet, but is a fascinating (terrifying) thought. So, a 1990s episode of *Tales from the Crypt* even presented this idea fictionally.

The idea of multiple universes and possible alternate realities and earths are popular fodder for comic storylines. DC Comics "Crisis On Infinite Earths" and "Flashpoint" storylines have all the heroes from multiple universes come together to save the multiverse. There's multiple Supermen from various Earths; think different versions of ourselves living another reality on another version of Earth.

This idea isn't lost on science, as physicist Leah Broussard is currently studying "mirror matter" to try to discover a parallel or mirror universe. Broussard believes that if you blast a subatomic beam through a tunnel and magnet, you could possibly have those particles "mirror" themselves and tunnel right through the wall. The theory says you could possibly find "mirror" particles, rocks, planets and stars. At the publication date of this book, the experiments have yet to take place. Zurab Berezhiani, a physicist at the University of L'Aquila in Italy, believes dark matter has been hard to find because it hides in the mirror world. This science exists in popular culture, as in DC Comics "Year Of The Villain" storyline the Batman Who Laughs is trying to infect Earth with dark matter in order for it to mirror his own earth from the dark multiverse.

The idea of other earths and fantastical beasts are common fictional tropes. If you look into popular culture, the Star Wars character Chewbacca is a Wookie, which is a large, bipedal, hairy, apelike figure. Sound familiar? Tales of Sasquatch sightings exist in various cultures from all over the world, and the description of large hairy giants appear in Sumerian and biblical texts as well. It wouldn't be too out there to suggest Chewbacca is based upon a cryptid-like Bigfoot. The kraken, giant squids, and the Loch Ness monster have all been reportedly sighted, and there have been fictional science fiction and horror movies mirroring these monsters of water. Mythological creatures exist all throughout recorded history.

In 2019, new cave art dated back 44,000 years in Indonesia depicted human-animal hybrid-like beasts called therianthropes hunting pigs and buffalo. An academic study into the drawings revealed much more.

"The hunters represented in the ancient rock art panel are simple figures with humanlike bodies, but they have been depicted with heads or other body parts like those from birds, reptiles and other faunal species endemic to Sulawesi (Indonesia)," said Adhi Agus Oktaviana to CNN.com (study co-author and a PhD student at Griffith University in Australia).

The journal *Nature* conducted a study on the sophisticated cave paintings that tell a story of supernatural beasts hunting prey, and considered the finding special since it displays examples of modern cognition like stenciling, storytelling and religious thinking.

"The images of therianthropes may also represent the earliest evidence for our capacity to conceive of things that do not exist in the natural world, a basic concept that underpins modern religion," said Adam Brumm (study co-author and associate professor at the Australian Research Centre for Human Evolution) in an interview with CNN. "Therianthropes occur in the folklore or narrative fiction of almost every modern society, and they are perceived as gods, spirits or ancestral beings in many religions worldwide."

If we look at Greek mythology for starters, we can compare these therianthropes in Indonesia cave drawings to a Minotaur. The Minotaur was a bull-headed monster born to Queen Pasiphae of Krete (Crete) after she coupled with a bull, and was born of the arrogance of man to the Gods.

Let's pretend not just a God, but various "Gods" exist. What does this say about all the tales of mythology?

Paranormal hot spots, high-energy areas, and other mysterious locations exist all over the world. The Bermuda Triangle is one of the original anomalies I looked at in *Punk rock and UFOs: Cryptozoology Meets Anarchy*, and looked at it as a possible wormhole or link to another dimension.

Theoretically, the idea of a wormhole would create passages through the universe that could connect space and time and serve as shortcuts to all over the universe. Think of it as a tunnel with a departure point and a destination at the end. Video-game nerds, remember those pipes in Super Mario Bros. that would send you to new levels? What about in Legend Of Zelda, *the flute that warps you from point A to point B? Well, a wormhole is sort of like that. What if the aliens have the power and technological prowess to travel through these wormholes and that is what makes their space exploration so seamless? It would make sense on how they can visit our* planet, yet we aren't advanced enough to travel to their solar system and galaxies. In a way, we as humans should feel an inferiority complex here. Almost like when your older brother is bigger and stronger and is able to muscle you out of the way for the last piece of pie. "Mom, it's not fair; how come they can come here and we can't go there!?"

There are lots of theories about these wormholes existing on Earth, with the Bermuda Triangle being the most famous. There is an aura of mystery surrounding this area of the Atlantic Ocean. More airships vanish in this region than most other areas of the world. When Christopher Columbus sailed through the area on his first voyage to the New World, he said he saw a great ball of fire that crashed into the sea. An odd light had developed days earlier in the same location. Paranormal enthusiasts cite Atlantis, sea monsters, aliens, reverse gravity fields, time-warp tunnels and the devil as causes for the Bermuda Triangle's ability to swallow planes and ships and to never return them. Where do they go? Are they sucked to the bottom of the ocean? Are they transported through time? Do they enter a wormhole to another universe?

Not every encounter with the Bermuda Triangle proved to be fatal. In 1970 pilot Bruce Gernon Jr. encountered the triangle, but lived to talk about it. On his *way to Palm Beach, Florida, from the Bahamas, Gernon Jr. approached the triangle area and was met with the clouds and fog that exist. Stuck in a cloud that resembled a tunnel, the passengers felt weightlessness as the craft picked up an unnatural speed and exited the cigar-shaped tunnel surrounded by a white and green sky, as opposed to the blue sky that was ahead. During this, the navigational equipment on board was not working, and he could not make contact through his radar control. Through the haze, Gernon Jr. spotted an island, which ended up being Miami Beach. He had burned 12 fewer gallons of fuel than usual and was to the destination of Florida 30 minutes off his usual time of 75 minutes. Somehow, this plane leapt through space and time. It's noteworthy that the cloud and tunnel were described as cigar shaped, as many UFOs are described in a similar fashion.*

The innocence of childhood, the wide-eyed wonders of eternal youth, and the feeling of nostalgia are common themes in popular works of literature, TV and films, both cult classics and smash hits.

Stephen King's short "The Body" became adapted into the film *Stand By Me*, which was a slight departure from his usual horror style. In *Stand By Me* the plot revolves around a group of young boys who find the body of a missing boy. King would revisit this idea of youth confronted with real conflict in *IT*, which has become one of his most popular stories in literature, TV and film. *IT* has a group of friends battling a shapeshifting creature from space (Pennywise the clown) that uses fear to prey on the children of Derry, Maine. The entity returns every 27 years to torture the children of the town, and in the second installment of the story the group ("the Losers Club") returns as adults to try to end Pennywise's reign of terror once and for all. An interesting plot point in Chapter 2 is that many of the adults have blocked out the horrific memories from their childhood and have "screen memories" to protect them from the trauma, which is a common experience in UFO abduction cases.

The 1980s cult-classic *Monster Squad* had a group of young horror-movie obsessed friends face their real fears as Dracula, the Wolfman, the Mummy, the Creature From the Black Lagoon, and other creatures arrive in their small town to wreak havoc. The group enlists the help of Frankenstein to help stop Dracula and his evil plans.

Borrowing from these ideas of nostalgia, small towns, paranormal oddities, and camaraderie among friends is the mega-hit Netflix show *Stranger Things*. While not reinventing the wheel creatively, the show introduces a new generation of people to this nostalgia factor and the eternal youth taking on out-of-this-world elements.

The fictional town of Hawkins holds some dark secrets, which include a facility that has CIA-run experiments on the character of Eleven and her telekinetic abilities as part of Project MkUltra. Eleven can also use remote viewing powers and access other dimensions. While doing this, Eleven encounters a creature living in the Upside Down dimension, and eventually opens a gate between the Hawkins Laboratory and the Upside Down dimension; thus the creature can travel between the human world and the Upside Down.

Does this sound familiar? It should. The CIA studies of remote viewing and using physics are well documented. Indigo Swann researched the process of remote viewing at the Stanford Research Institute and collaborated with Hal Puthoff, Russel Targ for a CIA-funded study. He is credited for proposing the idea of controlled remote viewing, which is defined as a process in which viewers would view a location given nothing but its geographical coordinates, and this was tested during the study. Uri Geller was another famous remote viewer who worked with the above team of scientists, who concluded that they did indeed have unique skills, despite many in the scientific community scoffing at these studies as pseudoscience.

According to Swann, he was able to remote view Jupiter and made several reports on the physical features of Jupiter, and previously unknown physical details ("rings") were later confirmed by a Voyager expedition. Swann also claimed to remote view a secret ET base on the hidden side of the Moon, and confirmed aliens live among us. There've been other projects that involved remote viewing on Mars and Project Star Gate, which were all CIA funded.

Now, back to Hawkins Laboratory, which forms an obvious parallel to Montauk, New York. Montauk Project was an alleged series of secret US government projects conducted at Camp Hero or Montauk Air Force Station in suburban New York for the purpose of developing psychological warfare and "exotic research" in topics like time travel, teleportation, mind control, contact with alien life, etc. UFO researcher Jacques Vallée looked into the Montauk Experiment stories that seem to have originated with the account of Preston Nichols. Nichols claimed to have recovered repressed memories of his own involvement and claims that he is periodically abducted to continue his participation against his will (sounds almost like an alien abduction?). The Montauk story is referenced in the fictional movie

Eternal Sunshine Of The Spotless Mind, and *Stranger Things* once tentatively had the word "Montauk" as a working title. The documentary filmmaker Christopher P. Garetano's film *The Montauk Chronicles* got mainstream attention from outlets like *The Huffington Post* and was eventually shown on History Channel.

Harry Potter remains one of the most successful franchises in fiction history with books, movies, theme park attractions, merchandise and fandom. J. K. Rowling's Harry Potter series deals with magic and wizards and was geared towards young adults. Introducing young adults to these stories fantastical in nature is no different than Hollywood making UFO films to help indoctrinate the public. These are "fictional" topics that appear in our movies and pop culture that were once believed, and currently still are believed, by some to hold some truth.

Netflix's *Chilling Adventures of Sabrina* is a darker reboot of *Sabrina the Teenage Witch* aimed at young adults that also deals with themes of dark arts, magic, witchcraft, religious history, mythology, and paganism. Mainstreaming these "controversial" ideas to youth through fictional series will only help open up their minds and maybe inspire them to research real stories related to these themes independently.

Urban legends and folklore are birthed out stories

believed to be true that circulate through communities and families throughout time. The iconic Scary Stories To Tell In The Dark book series are all fictional, short scary stories based upon urban legends. Urban legends vary from what could be considered jokes/pranks to a higher sense of plausibility based on true events. Does this mean every urban legend is inherently true or false? No, it means there are some truths to these tales, which is another example of the connections between fiction and nonfiction. *Grimms' Fairy Tales* could be considered one of the earliest forms of urban legends, folklore passed down via a "friend of a friend." The stories are being told and retold for a reason, which is to be determined by the world view you have on this. Are you convinced these are myths that help reinforce these weird and complex occurrences as true? Or are they simple stories meant to spook out your friends at a campfire?

There are some urban legends that cross over to the realm of cryptozoology. The chupacabra is a popular cryptid that many believe sucks the blood out of livestock. The Spanish breakdown of the name (*chupar* means to suck, and *cabra* means goat) means literal goat sucker. There's the paranormal creatures called the black-eyed kids that are described as pale kids with dark eyes, which resemble an alien-like figure that also could be described as how some of the Men In Black sightings appear. In Great Britain and Ireland there are tales of a nocturnal apparition that takes the form of a large menacing black dog. There's also sightings in North America of the dogman, a cryptid similar to Bigfoot sightings, but appears large, muscular in size, with the head of a dog. Sightings of a dogman have dated back to the 1800s. Tales of upright walking canines that look like a dog or wolf could also be associated with lycanthrope lore and werewolf stories. The Egyptian God Anubis is depicted as a human body with the head of a dog, which many witnesses describe a dogman as looking like. Large wolves that resemble the thought-to-be extinct dire wolf are a common sighting at Utah's Skinwalker Ranch.

Innovation birthed from a cosmic interference was the origin for this brilliant scientist, and his out-of-this-world abilities. His scientific acumen was strong as he was generally acknowledged as one of the smartest men in his universe. As a child he was a scientific prodigy with an aptitude for mechanics, physics and invention. Simply put, his work obsession with science would be considered fantastic.

– Reed Richards (Fantastic Four).

A mad scientist or a genius ahead of his time? A Dr. Frankenstein or a savior? Nikola Tesla was one of the great minds and engineers of all time. His work ranged from experimental to groundbreaking; his files were seized by the FBI upon his death.

Tesla's work involved his proposed "death ray," which harnessed a type of energy called "directed energy." "Directed energy" used powerful beams on excited electrons that could then be used on any target. UFO researchers believe aliens use this form of energy as part of the propulsion technologies of UFOs. Remember the stories of the God Zeus throwing lightning bolts from the sky in Greek mythology? Could this have been alien-engineered directed-energy technology at work?

What we now see as mythology, well, there could be truth to these tales. Mythology could very much be recorded history as it happened then, just as when we read about the Civil War in our history textbooks; we weren't there to live it, but we believe it happened. Now, why can't mythology be the recorded history of the time and we believe the stories from the Egyptians, Greeks, and early Sumerian civilizations are the recorded truth as they saw it or experienced?

Tesla believed his soul was full of too much energy. He sometimes slept just two hours a night and was haunted by visions and ideas that were "channeled" through him.

In 1919, Tesla wrote of these images and tried to find an explanation through several doctors and psychologists, but to no avail.

"The theory I have formulated is that the images were the result of a reflex action from the brain on the retina under great excitation. They certainly were not hallucinations, for in other respects I was normal and composed. To give an idea of my distress, suppose that I had witnessed a funeral or some such nerve wracking spectacle. Then, inevitably, in the stillness of the night, a vivid picture of the scene would thrust itself before my eyes and persist despite all my efforts to banish it. Sometimes it would even remain fixed in space though I pushed my hand through it."

Tesla also believed in extraterrestrials and spirit or soul that continues in another plane of existence after death, and life was just an "automation of nature." The "transmissions" he received in his head helped confirm his belief in the strange. Though he often denied he had any superhuman abilities, Tesla wrote:

"The sounds I am listening to every night at first appear to be human voices conversing back and forth in a language I cannot understand. I find it difficult to imagine that I am actually hearing real voices from

people not of this planet. There must be a more simple explanation that has so far eluded me. ... I am hearing more phrases in these transmissions that are definitely in English, French and German. If it were not for the fact that the frequencies I am monitoring are unusable for terrestrial radio stations, I would think that I am listening to people somewhere in the world talking to each other. This cannot be the case as these signals are coming from points in the sky above the Earth."

Elon Musk and Robert Bigelow both transcend the idea of the typical capitalist businessman. Their eccentric demeanor, interest in the unexplained, and use of their resources to push technology, space exploration and human advancement to the next level are Tesla-esque. Musk's electric car is named after Tesla. Musk, the CEO of aerospace innovators SpaceX, neurotech company Neuralink that looks to interface the human brain and computer capabilities, The Boring Company that focuses on tunnel building for electric vehicles, and other ventures all revolve around his idea of changing the world and humanity. His lofty goal of one day colonizing Mars also plays into his mission to use innovation to better human existence. Some of his innovations and creations look right out of a science fiction graphic novel.

Bigelow used his real-estate capital to start his aerospace empire Bigelow Aerospace, and wants to create the first commercial space station, which will include space labs, hotels, and factories. One of his innovations is the Bigelow Expandable Activity Module (BEAM). BEAM is an inflatable habitat (under contract to NASA) used for testing as a temporary module on the International Space Station (ISS) from 2016 to at least 2020. Bigelow was brave enough to declare on *60 Minutes* that he was "absolutely convinced" there have been extraterrestrial visitors to Earth. Bigelow is no stranger to the ET question, as his National Institute for Discovery Science (NIDS) looked into paranormal locations like Skinwalker Ranch, as well as the associated cattle mutilations and UFO sightings.

   "Synchronicity; an ever-present reality for those who have eyes to see." – Carl Jung

SYNCHRONICITY IS an experience of two or more events that occur in a meaningful manner, but can be unrelated. These events are usually random. This is a concept that was first introduced by analytical psychologist Carl Jung, which holds that events are "meaningful coincidences" if they occur with no causal relationship yet seem to be meaningfully related.

In *Punk rock and UFOs: True Believers* I used the quote from Jim Gordon in *The Dark Knight Rises* of "You're a detective now, son; You're not allowed to believe in coincidence anymore," to connect the reader to the idea of synchronicities without implying its definition. Besides the possibility of synchronicities, there are also commonalities that exist between certain unexplained elements.

The shape of a triangle appears multiple times in UFO lore and history. First, there are triangular-shaped UFO crafts, which are one of the common sightings.

A pyramid is a geometric structure whose outer surfaces are triangular and meet at a single step at the top. The Egyptians built the pyramids as tombs for their pharaohs and believed that when the pharaoh died, his spirit remained vital in the afterlife. The pyramid also serves as an integration of self and soul. The ancient Egyptians saw the shape of the pyramids as a method of providing new life to the dead, because the pyramid represented the form of the physical body emerging from the Earth and ascending towards the light of the sun. Egypt isn't the only place pyramids show up in history, as the Mayans built many in Mexico, and other pyramid-like structures have been discovered all around the world. It's quite an anomaly that unconnected civilizations all over the world decided to build the same structures. The Washington monument in DC also features a pyramid shape at the top.

From *Punk rock and UFOs: True Believers*:

The top of the Washington Monument, built by Masons, also forms a triangle at the top, almost *appearing as a mini pyramid. The Pythagorean theorem ($a^2 + b^2 = C^2$), popular in Mason mechanics, also forms a triangle. Coincidence? Maybe. In mathematics, the Pythagorean theorem, also known as Pythagoras' theorem, is a relation in Euclidean geometry among the three sides of a right triangle. It states that the square of the hypotenuse is equal to the sum of the squares of the other two sides. Numerology and mathematics were huge parts in the functioning of early civilizations all over the globe, as they had an advanced understanding of calculations and the practical application of mathematics into their infrastructures, advancements and daily life. ... Masonic lore is full of tales of doom and apocalyptic prophecy. George Washington, one of the most iconic Masons, was said to be visited by an "angel" or higher celestial being warning him of future wars in America and upcoming turmoil facing the nation. Through Mason's mathematics and architecture they created a calendar—aligning the great pyramids in 1473, Solomon's Temple, and the Washington Monument reveals a doomsday clock with the tip of the Washington Monument proving to be the end. The prophecy is fulfilled in 2022; Armageddon or end of days when God comes down for judgment. Most cultures and religions hold a prophecy of the end of days and when a higher power returns to earth for judgment. This isn't a* *mutually exclusive idea of just the Mayans and the Masons.*

Then you have the aforementioned Bermuda Triangle, which is a triangle-shaped anomaly that led to the disappearance of those brave enough to try to go through it, with one survivor noting he went through a cigar-shaped (there are cigar-shaped UFOs as well) "portal."

Experiencer Christopher Bledsoe has claimed visitations his whole life. His stories do note interactions with orbs, UFOs, and aliens that take place in our physical realm and not just in his consciousness. He has claimed to see them in the flesh, and one of his stories notes one of the beings had a triangle symbol on the chest of its uniform.

Bledsoe has also seen beings that glow like a light, which was the same color glow as some of the UFOs he witnessed, which begs the question, do UFOS and the aliens that inhabit them all manifest from the same energy?

Bledsoe's interactions with these beings in broad consciousness align with hundreds of years' worth of reported sightings of similar nature. Today, these sightings seem rare, but there are hundreds of reported interactions that appear in religious texts, Native American lore, and news reports. In many cases, beings appeared to be witnesses in our physical realm, which differs from the abduction cases usually involving the greys. Some of the reported interactions throughout with beings that were witnessed outside or exiting what we would call a UFO share some things in common. The various reports described some of the experiences with the inhabitants as conversational. In the 1800s many witness reports described what we would think of a primitive UFO or spaceship – something out of a steampunk novel comes to mind. The inhabitants were human in appearance and in many cases were stopping for supplies like water or other elements and conversed with the witnesses. There was something that they needed for the ship, and at times they would reveal where they would go next. In some cases UFOs and the inhabitants arrived on farms and left with cows or chickens, which could easily be a cattle mutilation-type case. Some of the other descriptions of beings included "dwarfs," "frog-like faces," humanoid shapes with dark skin or pale skin, tall humanlike figures with long hair and slanted eyes (Nordics?), and in many cases they were all wearing some tight-fitting uniform or overalls that could also be compared to a ski suit. They also had "red hat"-like helmets, and helmets that were also described as diving gear. Some of the beings have been said to carry a light rod that can emit energy almost like a magician's wand.

There are fictional equivalencies to these real-life

sightings in science fiction pop serials like Ufonauts, dwarf/elf/fairy folklore, and the idea of time-traveling humans. Even the UFO itself and its various shapes and sizes, along with various reported races, is sci-fi fodder when you look at the universe created in *Star Wars* and *Star Trek*, only UFOs aren't fiction.

There's science fiction and speculative fiction. Science fiction gives more plausibility to fantasy stories by incorporating technology as opposed to "magic" you'd see in fairy-tale stories, which ultimately makes the story more "believable" to audiences. Those interested in UFO studies are also drawn to science fiction.

"Humans love a good mystery," said Rogue Planet TV's Jason McClellan. "We're fascinated by the unknown. One only needs to look at the explosion of the True Crime genre to realize the widespread public interest in mystery. Humans are curious creatures, and we look for answers to explain mysteries and rationalize unexplained phenomena. Science fiction is a genre where supernatural themes and mysteries on a universal scale are common. Science fiction steps outside of our reality, and it explores many of the grand mysteries we contemplate every day. Because space and extraterrestrial life are regular themes in the Science fiction genre, it stands to reason that some of those who were drawn to this genre because of their curious nature and a proclivity for the unknown eventually take this casual interest beyond passive entertainment consumption and venture into researching topics like UFOs and other unexplained phenomena."

Speculative fiction is where the author speculates upon the results of changing what's real or possible. Whatever is being speculated on must be larger than character or the plot. *Blade Runner, Jurassic Park, 2001: A Space Odyssey,* and *A Nightmare on Elm Street* are all considered speculative fiction.

Speculative fiction does a good job of asking the audience "what if?" "What if?" is something we ask ourselves all the time when thinking about the future or potential scenarios based upon decisions humans make daily. When studying the paranormal and inserting new and "far-out" ideas, an open mind may find itself asking the same question of "what if?" when presented with information or ideas that fly in the face of convention.

"Paranormal stories inherently lend themselves to this exploration," said two-time Bram Stoker Award nominated author of speculative fiction/horror stories Annie Neugebauer. "The nature of the results – fear, awe, intimidation, and so on – ultimately decides if paranormal works fall under horror, science fiction, etc."

There is room for some flexibility and fluidity in speculative fiction since technology changes what we know as possible. With technological advances, the belief from fiction to fact shifts, as does our preconceived notions of what is possible in the first place.

Student archivists at Rice University in Houston have been compiling, digitizing, transcribing and organizing massive archives from the works of Jacques Vallee, Whitley Strieber, Ed May (physicist contracted by CIA during the Stargate program), Richard Haines (ex-NASA scientist), and Larry Bryant. The spoon allegedly bent by the mind of Uri Gellar during CIA experiments is just one of many important objects being housed at Rice. (Photo by Mike Damante.)

*Chris Bledsoe's painting of a being he claims he was
visited by. Take note of the triangle on its chest.
(Photo courtesy of Chris Bledsoe.)*

Mike Damante gives Tom DeLonge a copy of
Punk rock and UFOs: True Believers *during the*
Angels & Airwaves 2019 stop in Austin, Texas.
Former Lockheed Martin/Skunk Works director
of Advanced Systems Development and current
To The Stars Academy of Arts and Science
member Steve Justice was also in attendance that
night. (Photo provided by Mike Damante.)

Leslie Kean with former Clinton administration Chief of Staff John Podesta. Podesta provided the forward for Kean's book UFOs: Generals, Pilots, and Government Officials Go On Record *and has been an advocate for UFOs and FOIA. (Photo provided by Leslie Kean.)*

66 *"Myths lose their power if they aren't repeated" – Batman: No Man's Land*

THERE WAS *a brilliant professor with the doctor distinction attached to their name. This doctor was intrigued that there were others out there with exceptional abilities, and needed to come together for a common cause. The doctor knew that these exceptional people were hiding in the shadows for fear of being discovered. Some of these exceptional people were common folk, while others were hiding under their elite status. What if the public would find out about their alleged abilities? How would they be perceived in their social circles? Would the government come after them? Would the public vilify them for being different? The doctor knew these were some of the obsta-*

cles in the way of the ultimate journey that could bring people together, but many continued to want to stay "invisible."

— *Professor Charles Xavier*

Diana Walsh Pasulka's pedigree is already impressive as a professor of religious studies at the University of North Carolina, Wilmington, and chair of the Department of Philosophy and Religion, and has served as a consultant on the popular The Conjuring film franchise. Pasulka's focus leans on the synergy between religion and technology, and that process of belief powers *American Cosmic* by tying in her previous studies of the supernatural and tech to the UFO question.

American Cosmic almost feels like a fiction novel at times, as her story is told in a narrative form that takes the writer all over the world with a supporting cast of real-life characters that seem straight out of a political thriller. Traditionalists may scoff at the more paranormal elements, but these viewpoints are slowly taking over the conversation in UFO studies, and *American Cosmic* is a well-researched example of this. The book talks a lot about consciousness, our DNA as a satellite, remote viewing and other anomalous abilities that are becoming more "mainstreamed" ideas into the UFO community with all the recent studies and experiencers coming out.

"In the field of Ufology, there is a traditional split between those who want to study the phenomena from a completely non-religious, non-woo-woo, nuts and bolts, 'scientific' perspective," said Pasulka. "Then there are those who feel they are star seeds and are souls here to help humanity evolve, and this is more of a spiritual or religious-like perspective. What I found was that among the scientists who I talked with, who studied the 'objective' parts or artifacts, the woo-woo aspect was quite prevalent. In other words, the scientists, though they would not refer to themselves as star seeds, were interested in identifying how individuals can have information that cannot be accounted for in any traditional way. In the beginning of the research I looked into the Russian and American Space programs, and again, found a lot of 'woo woo' there, in the history of the creation of rocket technologies. Jack Parsons, with whom your readers may be familiar, is just one example of many. He is the most well known because he was so flamboyant, quite photogenic, and had a lot of infamous friends, and tragically, he died very young."

In *American Cosmic* Pasulka meets an interesting man called "Tyler D," who has an air of autonomy around him in the book. The author alludes to him being a well-known MMA fighter and a highly successful tech entrepreneur. "Tyler D" could be one of many others in high positions of power and wealth like him with his perceived abilities.

"It took me a while to figure out that there were a group of these super high functioning individuals who are 'not known,' and I coined the term 'the Invisibles' to reference them," Pasulka said. "They work under the radar, but they are highly influential. He is one example from an elite group. However, if you met him, you would not, in any way, guess that he actually does the things he does. So, yes, there are others like Tyler, and although I met a few, I don't know how many there are. Tyler has an unusual ability to predict the future, which explains in part his improbable success. He credits his 'protocol' with allowing him to predict outcomes, and maybe this is true, I don't know. His protocol involves physical and mental exercises. Before I met Tyler and people like him, I assumed that many people with his level of success could be Googled, or would have a digital foot-print. I found out that, indeed, there are people who are very successful, influential, and are not 'known' in the 1st century way of being known—that is, by looking them up on the internet. To us, they are invisible. Thus, in effect, for us they don't exist to our conscious minds, but we are certainly impacted by the technologies they invent and produce."

One big revelation of the book is comparing Tyler's experience to that of the bilocation case that, oddly enough, both "took them" to New Mexico. The odd synchronicity makes it feel like the story *American Cosmic* is trying to tell really comes full circle.

"This is a good question," Pasulka said. "Frankly, I felt that the book took on a life of its own and wrote itself. I thought I was finished with it after Rey Hernandez's story. The book was way over its schedule (by a year), and when I thought it was finished, I was on a trip to the Vatican. About halfway through my trip, which included a trip to the Space Observatory in Castel Gandolfo, I realized that this was going to be my last chapter. Basically, events came together that really had nothing initially to do with each other, but somehow seemed to be connected. I was asked to go to the Vatican Archive to look through the canonization records of the seventeenth-century levitating priest, St. Joseph of Cupertino. I was also going to look through the records of Sr. Maria of Agreda, a Spanish nun who, around the same era as Joseph, was alleged to bilocate to New Mexico. And, I was with Tyler D. He is an expert on aerial phenomena. It should be noted here that in the beginning of the book I take a trip with Tyler and James, another scientist, to New Mexico to an alleged UFO crash site. So, at the Space Observatory archive, sitting there looking through books by Johannes Kepler, I suddenly realized that Sr. Maria said she 'astral traveled' to the place where I may have been. That struck me as really weird and eerie, whether or not it actually 'objectively' happened. So I mentioned it to Tyler, and he was also shocked. At this point I had become aware of the astral travel of people who believed that they 'remote viewed,' and that some of them believed they saw extraterrestrials, planets, and that they traveled through space. A lot of 'experiencers' who I had met told me similar things. Given my work in the history of Catholicism and Christianity, I happened to know that this is, though not common, something that others have reported to have done, like Emanuel Swedenborg, a contemporary of Immanuel Kant, and also a very smart person (said to have one of the highest IQs in history). This event helped me place Tyler within this history of exploration and, strangely, colonialism. Sr. Maria's experience was used by the Spanish government and Church to justify further colonization of that area of the New World, during that time period. She didn't authorize this, but that is how her narrative was used. In a sense, she was a pawn. And, in many ways, Tyler was a person whose own work was being used to further space research, and in a sense, he is being used just like Sr. Maria was used. He agrees to this, of course, so he has more agency than Maria, but it seems to take its toll on him."

American Cosmic also re-introduces the idea of the "book encounter," which we've seen in cases like Whitley Strieber and others. If the idea of the "book encounter" is truly a channeled guide, and people are being led to certain books specified for them, and for those who author books is it possible their area of concentration and appeal is also being manipulated?

"The book encounter is the event in which you find that you need specific information, and then voila, a book, or movie, or song suddenly appears and provides that information, or meaning," Pasulka said. "It is an uncanny experience that I don't understand, but that I often have, and others do too, so much so that some scholars have theorized about the phenomenon. Arthur Koestler, who wrote about coincidences, called this event the 'library angel.' I noticed that all of the experiencers, and invisibles, and meta-experiencers, had copious book, movie, and song encounters, so I needed to write about it. I can't say that I understand it other than it being a very meaningful and strangely timed coincidence, but I do offer an example of my own book encounter with *The Gay Science*, (which is) a wonderful book by the philosopher Friedrich Nietzsche. After my book was sent off to the press, I read Eric Wargo's book called *Time Loops*, which he theorizes time, coincidences, and precognition. For him, these kinds of events involve our future selves or minds, looping backward and providing us with the needed information. So, that is his provocative and interesting explanation. His book is very interesting, and he readily admits he cannot prove his theory, but he does provide a lot of fascinating examples that offer an explanatory framework for coincidences and synchronicities. Joseph McMoneagle, a person known for his remote viewing skills, mentioned something along these lines as well, in one of his online lectures. My take-away from all of this is that there is something about coincidences and synchronicities that invite us to ponder, take us aback, inspire awe. If anything, that is a good thing. If there is more, as Wargo and others surmise, than I think that they provide a language of sorts. If I do write another book about this topic, I will focus on this: the idea of synchronicities as a language."

I was going through some old boxes one day, and I found a literary magazine I appeared in as a child. The fictional story I wrote was about an alien abduction, which struck me as odd as an adult because I didn't recall being privy to any of this at a young age. For myself, the idea of a "book encounter" didn't come to me naturally, but my foray into the fringe came from a book subscription series by Time Life that included books on UFO phenomena, strange places and mysterious creatures.

From a philosophical perspective, what does this "wacky" idea of a book encounter say about UFO researchers and paranormal authors? Does this mean we are drawn to these books based on a cultural predisposition or an outside intervention?

"It shows our diversity, as a community, but also as humans," Terra Obscura's MJ Banias said. "We are all drawn to different aspects of research and, more broadly, reality. It depends on who you read, but 'book encounters' could occur for a whole host of reasons. Some could probably be explained via social and cultural upbringing (nurture), by some innate predisposition (nature), and, to get spooky, perhaps by some yet unknown force, like divine intervention, which draws us to specific ideas and texts. Whatever the case, it complicates the UFO research matter. We all approach UFOs and, more broadly, the 'paranormal' from different angles. There is a creative and chaotic beauty to this, as it creates an incredibly democratic and even anarchic system. No or knows what the phenomenon is, and no one can really answer the big question. This has obvious problems; it remains a mystery and no one has made any progress in getting a better understanding as to how this mystery plays out. We are chasing ghosts."

There is an easy opportunity for the lines to be "blurred" when doing fiction and nonfiction, but it appears that throughout history nonfiction and fiction have met in the middle at various times. The occult, tales of magic, beasts, giants, angelic beings, death days, and Gods are all themes that have weaved throughout stories and reported history.

There's been a valid argument made that through years of Hollywood productions, the lines between fiction and nonfiction are blurred. Some argue it's a bad thing that gets in the way of proper disclosure, while others see it as a necessary evil in making the public aware of potential out-of-this-world scenarios. If you go back to the earliest works of fiction and recorded texts, the lines were blurred from the start. Stories are usually based on something, and as vast as our imagination is, the world around us influences creativity greatly. Our imagination is influenced by reality, and our imagination helps better our reality.

From the beginning, texts and writings were influenced by what the scribes and authors saw and believed. Stories of dragons, wizards, flying chariots of fire, dwarves, and giants didn't just manifest out of nowhere.

If we examine various avenues and genres of fiction, you can trace a real-life counterpart thematically somewhere to link them both. The superhero genre is rich in the history of humans, mutants, and beings with extraordinary powers who are able to pull off impossible feats, but how is this different from the powers that the Gods had in Greek and Egyptian mythology? Science fiction tells tales of space exploration, space wars, multiple races of aliens in space interacting through space and time, time travel, alternate realities, robots, and other "weird science" that isn't so weird when you look at the breakthroughs we have seen throughout the years in technology, medical and scholarly studies, and centuries' worth of reported sightings of UFOs and interactions with otherworldly beings. Medieval literature tells tales of dragons, witches, and sorcery, but for the European cultures who've reported for years fauns, fairies, elves and other anomalies, the folklore becomes legend.

If these possibilities exist in a universe full of infinite possibilities, is it too far-fetched to believe at some poin giants walked the earth, magic was real, and humans possessed powers that the Gods once did?

For those advocating keeping fiction and nonfiction separate, it is too late. The lines have been blurred from the beginning. Being able to differentiate between fact and fiction is still a vital tool in our overall intellectual discourse, but let's remind ourselves that every time we go to a movie and see big Hollywood blockbusters like *Indiana Jones* or *Star Wars*, they are rooted in some form of reality. The idea that Hollywood, through the encouragement of the government, has made UFO films to help ease the public to the possibility of one day knowing the truth has merit. If we think about other themes and narratives Hollywood does, well, you can apply the Hollywood-helping-the-disclosure-movement theory to various other works as well. Do films about reincarnation, time travel and monsters mean that they are intended to one day "soften the blow," or is it more than just a reflection of our modern-times way of visual storytelling?

The capacity to see the impact fictional literature, stories and film/television have on our society is important. Fandom for some is a way of life. Comic Cons, fan fests and other conventions help bring senses of community, and these fictional characters inspire real people to do better and help the world around them. This is *hope*.

The universes created through fictional franchises offer more than just escapism, but a parallel world where possibilities some may not be able to experience can live.

We can't fault the way humanity has advanced, and how science and factual accuracy are cornerstones of society. Imagine where we would be without the advancements of science, the freedom of the press, and an emphasis on rational thinking and education? At some point in history though, things *shifted*. If we take at face value the sensational stories in religious texts and early recorded tablets of the world's first civilizations, we would accept it as facts/recorded history. History as they saw it happen. Nowadays we think of the idea of giants, angels/demons and UFOs as unproven or some may say outlandish, but hundreds of years ago life on earth was different than we know today. Who are we to scoff at past societies' stories and deem them as just "stories"?

Many are dedicated to finding the facts, searching for the truth, and practicing general journalistic standards in the study of all things UFO, but somewhere along the way many authors switch their focus from nonfiction to fiction. Why is it that so many immersed in the field of UFO research are drawn into the world of fictional writing?

"Because of the dizzying and sometimes fantastical nature of UFO phenomena, it seems like some sort of mystical marriage between those who are interested in UFOs and those who weave imaginative stories in their heads for books, film, and television," said Ryan Sprague from CW's *Mysteries Decoded* and host of his own podcast *Somewhere In The Skies* of the same name of his book. "The lines between fiction and truth are blurred to such an extent merely by the way we process, interpret, and tell our stories. Both seem to influence one another in such a subconscious way that the alien and the UFO are *us*. They always have been."

There's an obvious parallel between science and science fiction; at times the lines of fiction and nonfiction are blurred. Many believe science fiction is based on true-to-life events or has some basis. Or has Hollywood infiltrated our collective consciousness so much that we are inclined to believe what sounds unbelievable to many is believable to us?

Were films like *Close Encounters* and *E.T.* based on real events and used as part of a slow acclimation of the idea of aliens to the public? Hollywood has been spreading UFO awareness through the years through works of fiction, as UFOs, aliens and all things outer space have become integral parts of pop culture. Films about UFOs were allegedly encouraged by the government to likely get the public's mind off the real sightings that were increasing. The public was to be entertained and this was "camouflage through limited disclosure" as described in the book *The Day After Roswell* by former Pentagon official Colonel Phillip J. Corso, who worked closely on US government projects regarding reverse-engineering alien technology from the Roswell crash.

Whitley Strieber was a successful fiction writer before publishing his nonfiction abduction account *Communion*. Strieber even pondered in his books if his horror fiction writing was influenced by his repressed memories, emotions and fears that stemmed from being an experiencer.

Have we evolved as a society to what once was prevalent on earth is now hidden? If other worlds exist on Earth, they will remain hidden, as mainstream interests see little need to chase "fairytales." What if other worlds (Hollow Earth theory), parallel universes, and wormholes to other dimensions all exist in our world, but are hidden from our naked eyes to see? If only certain selected or enlightened people, or those who are able to navigate through their own consciousness, are able to see these hidden worlds, well, what does that say about those people themselves? Are they magic? Are they operating like the Gods? Are they using a part of their brain or DNA that most don't use, and remains dormant for years?

When we accept that reality isn't black and whi

and in those shades of gray where fiction and nonfiction intercept, we can open our minds to possibilities that go beyond what we've been engrained to believe can only exist in the reality we've created. If you look at the world's greatest thinkers and innovators, their ideas stem from their imagination and were previously thought to be impossible. When we can make the impossible possible, we can enrich the world around us. Science and science fiction can coexist, and we can see fictional stories as something other than just make-believe; we can dig deeper to the origins of those stories, and the historical allusions that accompany them, and better learn about our history and ourselves.

There's this idea that the "powers that be" (world leaders, CIA, religious leaders) play a role in keeping some of these possible truths at bay for fear of public outcry, failure of our institutions, and various reasons that stem from nefarious conspiracy to moderate secrecy. If this holds true, for the most part they are winning. We tend to question things that seem possible rather than those that are unfathomable. If the ideas that UFOs can't possibly be real exist, then any secret government program to block disclosure has done its job since we are still having this debate today. Fictional works are just that, fiction, but these same people in power will tell you the Bible is true. The dilemma we face, which is an uphill battle when you think about how marginalized non-mainstream ideas have been in the face of traditional science and traditional belief, is resurrecting past possibilities as feasible today. We've buried our "Gods" in exchange for smartphone screens, our hidden worlds remain just that, and the UFO debate continues to divide a community born of toxic skepticism and existential exhaustion.

If there is truth to past mythology and legends from folklore, maybe they never really went away, but they "live" today through fictional depictions and stories.

Now one thing this book is not advocating is a total abandonment of reality or believing everything you read in fiction as truth. We still need to strive for accuracy and a grip of what is real, but we need to expand our minds to the possibilities that exist beyond just what we've been told is possible.

How do we expand our minds to these themes that seem too fictional to be true, but at the same time grasp what is factual and possible as we know it? Where exactly did the "paranormal" stories that existed in everyday life in religious and mythological texts fade away, and were replaced with what we'd call today a more "possible" truth?

"Everything changed," said professor of religious studies at Rice University Jeffrey Kripal. "Scien happened. Technology happened. Religious pluralism happened. It is not very difficult to believe anything religious, right? I don't think paranormal experiences are simply 'true' either, by the way. I think they inhabit some strange space between truth and fiction."

*T*HE FOLLOWING IS INCLUDED *as a small chapter to help tie in the themes together. The below is an account from a dream I had before I even started working on the book. It is here to provide context to an implied greater journey we may all be encountering.*

I had a very vivid dream that I was on a tour of the universe. I'm not trying to make any assertions or claims of channeling or anything of that nature here, but I just found a lot of the circumstances and occurrences of the dream interesting and oddly specific. Below is an account of what I remember.

I had a tour guide during this "trip" to the universe in the dream. The tour guide was a lady, but I felt it was something just taking the form of a middle-aged librarian-like woman. At one point all that could be seen was space, which was illuminated by many red stars, and I was told by the tour guide, "Don't be alarmed; it is beautiful."

Then the tour guide took me into what appeared to be a candy or gift shop, but the whole time I felt in the dream it was a shield or a screen for something else. While there, I had to wear a white hard hat so I wouldn't be recognized as being "out of place."

This is where it gets interesting. In the dream, I gave the tour guide a summary of a theory I proposed in my first book, *Punk rock and UFOs: Cryptozoology meets Anarchy*. The theory from the book:

"What if our souls were a simple light force or consciousness? What if when we die, our 'souls' go up past the ether into the atmosphere and find its way into another universe or a 'heaven'? What if heaven is really another universe that can only be traveled to by our consciousness that contains our memories and cognition?"

The tour guide's response to this was "give or take." I can't recall specifics about other questions I asked the guide, but I remember all the responses being short one-to-two-word answers. What is striking about this is from what I've read about hypnosis/regression cases where patients are claiming to channel responses from an unknown voice and are asked specific questions about the universe or phenomena, the brief answers were similar to the "give or take" reply I got in the dream. I hadn't read or studied a regression case deeply in over a year.

In the dream I also asked about the greys, and the tour guide quickly changed form to a grey alien and back to human form. After that the guide was showing me candy, and a human lady walked past us. The guide informed me that it was "Dana Hickenlooper." When I woke up, I looked up a "Dana Hickenlooper," but nothing really came up. The last name Hickenlooper rang a bell because Colorado Governor John Hickenlooper was running for president at some point. Wild speculation here, but could he possibly be one of the candidates in the know of secret UFO programs?

At the end of the day this was just a dream, but I felt compelled to share because I found some of the specifics and imagery to be oddly coincidental at best. I'll leave you with a small passage about dreams from *Punk rock and UFOs: Cryptozoology Meets Anarchy.*

Have you ever had a very vivid dream that just seems like it was real? Have you ever had a dream you were drowning and you find yourself forcing yourself to wake up and gasping for air when actually awake? I've had a recurring dream that I was being attacked by a shark, and I always remember forcing myself to wake up, and when I do, I am out of breath and my ribs feel sore. Night terrors are sleep disruptions that are more severe than just a nightmare. Sleep occurs in several stages that correspond with a function of the brain. Most of our dreaming occurs during the rapid eye movement, or REM, part of sleep. A lot of nightmares happen during the REM stage, but a night terror is a reaction to fear that occurs during transitions from one sleep stage to another rather than an actual nightmare. Night terrors can be caused by an active or overly aroused central nervous system (CNS). Night terrors are considered to be a parasomnia, or an undesired interference, during sleep. Abductions and visitations can be called parasomnia too, since many occur during our intimate and fragile sleep state. There are similarities made from a nighttime visitation and night terrors in how helpless and how there is a state of uneasiness during both.

Lucid dreams are dreams where the dreamer is aware they are in a dream and aware of self, function, decisions, meaning, environment and overall clarity.

Fiction part: Two short stories

1. **A Gift**

2. **The Cosmic Trickster**

A Gift

Downtown always lit up the sky at night, but not like this.

The bright lights of the big city were just as bright as the stars in the sky and would stay illuminated throughout the night even when the city slept.

Bustling with nightlife, traffic and energy, the city was the epicenter for commerce, social gatherings, dining, and entertainment. Usually when the bars closed down late, the city would die down too, but the lights remained on, as from the suburbs you could still see downtown shining.

That was why it was a shock to many what occurred one curious morning. In an undeveloped part of downtown where no one lived, and just a few old warehouses stood, one morning a giant structure now stood in its place. The structures were big and, like trees, grew from the ground up towards the neighboring buildings and skyscrapers. Protruding up from the ground were stone-like formations that looked like tree roots, with some forming a triangle shape at the top. Some of the structures almost looked like the Washington Monument, as they grew straight up and were pointed at the top, where a triangle formed. Others were more vine-like and inter-woven around the other formations and almost formed a tree at the top, just with no greenery.

That morning the inhabitants of the town gazed in awe at the monoliths that hadn't been there a day prior. Soon news spread of this, as news crews were on the scene, speculating about this mysterious happening. No one downtown said they had seen anything the night before. Even the city's homeless population didn't come forward with anything on how this had been built.

A few people in the suburbs said they had seen what looked like a shooting star fly over downtown the night before. One of those witnesses just happened to be a man named Phil Reed. Reed had been working air traffic control at the airport, which was a good 30 miles outside downtown. He said he saw a shooting star crash down-town. No one saw it crash-land or touch down, and the people who lived downtown didn't hear anything. A few residents mentioned a late-night disturbance, saying they felt a small shake in the apartment, but dismissed it as maybe just a truck crashing.

Since the structure arrived, downtown was brighter than ever. It was lit up before, but now the city had a glow to it. The structure had given this city an aurora to it, almost feeling and looking like a city from the future. Scientists tried to figure out the properties of the stone. It was almost impossible to get a sample, but preliminary test results came back inconclusive. The elements of the stone were not of this world. Some debunkers saw it as an elaborate hoax, or an "art project" that the city and some local architecture had partnered up to create to generate interest and tourism. Religious leaders from all over the world questioned what it meant. Some UFO enthusiasts claimed it was soft disclosure.

The structure that was the downtown anomaly remained a mystery. Why was it here? What was it? Where did it come from? If it was anything, it was a gift.

The Cosmic Trickster

The galactic federation held their timely meetings in a galaxy far, far away, but not far enough for one. The cosmic trickster hated these meetings. It simply wanted to do what it did best, which was to distort realities and prank all over the universe. The cosmic trickster didn't care about bureaucracy, or what was best for the universe and the various federations that ran it. It simply just wanted to have fun.

As the cigar-shaped craft roughly around the size of a football field floated through time and space, the inhabitants of this meeting pulled up to the table. This meeting would be smaller than usual, as the greys and some of the other delegations were off for this meeting. The meeting room was silver and sterile. Just a long meeting table in the middle and a small window on the craft's wall to look out into the universe. This meeting would be led by Captain Zakari of Earth-8 and would also include Thoroson representing the Nordics, Echidna of the Reptilian race, and finally the cosmic prankster, Iocus.

"Where is everyone else today?" Iocus asked, looking around the table that had many vacant seats.

"This won't take long, and not everyone in the federation's participation is necessary," said Zakari.

"Good." Iocus sighed. "Because old snake face over here is freaking me out."

Echidna scoffed at him as she rolled her eyes up at the ceiling.

"Trust me, I'd rather be on planet Earth disguised as a human and interfering with their democracy," Echidna said.

"Show some respect, you fool, and bow down to the lady!" demanded Thoroson, slamming his large white hand down on the table and looking seriously at Iocus.

"Order, order!" proclaimed Zakari.

"I'm sorry for that one time I turned your lightning into liquid, Thoroson," Iocus said. "Come on, it was funny. Even the Alphans thought it was funny, and they don't laugh at anything. Water under the bridge?"

Thoroson stood up. His huge frame towered over everyone at the table, and he peered outside, looking into the vast, bright universe. The ship was surrounded by bright stars and glittering galaxy lights.

"The reason we have called this meeting is actually about your antics," Zakari started. "Iocus, you are not in trouble, and this pains me to say this, but we actually need you to do more of your trickery."

"Oh, do tell," Iocus said.

"The job is Earth. The original one ..." Zakari started.

"You mean that place where the humans are destroying their own planet?" asked Iocus.

"Yes, that dump," Echidna chimed in.

"Yes, Iocus," began Zakari. "It seems there are some

who are starting to get a little too privy to what is going on with your elves. The popular opinion polls in Ireland now state the majority believe they exist. They need to stop making themselves visible. One in particular ran a prank where in broad daylight it snatched a lollipop from a child."

"Amateur work – I'll dock the pay of the elf in question," Iocus said. "What else you got?"

"Well, the UFO disclosure movement is picking up steam too," Zakari started. "There seems to be a large contingent of people slowly putting this all together. I need you to go there and place a little disbelief in their reality."

"The humans get all hung up on words, letters, and titles ... I know just what to do, but I'm going to need permission to alter some timelines ..." Iocus said.

"Since when do you ask for permission?" Thoroson asked.

"Well, last time you did it with the Berenstain Bears books; that was pretty funny, I'll admit," Zakari said. "Do what you need to do ..."

Iocus in a flash appeared on Earth – the original one once populated by the dinosaurs. His destination was the United States of America, where the people were usually the easiest to prank. The Pentagon had secretly funded UFO studies, and the reports made front-page news of the *New York Times* and other mainstream publications. There were two specific programs: Advanced Aerospace Threat Identification Program (AATIP) and Advanced Aerospace Weapons Systems Application Program (AAWSAP). Both were causing division among the UFO community, and questions arose if the people who claimed to run it really did. Iocus loved this division and laughed as the humans argued on Twitter over these details while slowing the progress of their own advancements. Iocus decided to make things a little harder. With a cosmic snap of his synapses, he was able to slightly change history to now have it exist as the Advanced Aerospace Systems Study (AASS). But those who studied it still remembered it as AATIP, which caused more of an uproar. The slight altering of reality was all the cosmic trickster needed to do.

"Was it AASS or AATIP – someone is lying, and if the Pentagon doesn't give me what I want, then all hell will break loose," tweeted @TheWhiteCellar.

"It was neither. It was AAWSAP!" tweeted @UFOYoungGunNo1.

Iocus was finished with the mission, but on his way out, he happened to pass by Ohio and sprinkle some trickster magic over documentary filmmakers trying to find goblins in rural Kentucky.

The cosmic trickster's work was done. For now ...

. . .

BONUS Chapter: A real-life Captain America

I wanted to include this portion and story in the book, but I felt like parts of what is presented below may be a little too fantastical for some to swallow. I was able to get some character confirmations and check the validity of some of the documents that were provided to me.

There are parts of his story that aren't too "out there" for the UFO crowd, there are parts that governmental conspiracy enthusiasts will eat up, and other parts that really stretch the line of rationality, but the book is called *Stranger Than Fiction*, so it was apt to include it. This isn't an endorsement of his story, but like all good tales, it deserves to be heard – fact or fiction.

All he ever wanted was to make a difference, but he was simply not big enough, strong enough or fast enough. The kid had heart, which was evident by his continual tries to enlist, and serve his country. The boy's heart could potentially be exploited. What could go wrong? If anything happened, no one would miss him, would they? There was an arms race building up among other nations – the

Germans and the Soviets seemed to be on the cutting edge, so what could the US do to combat their advances?

The military had its guinea pig, and boy was that guinea pig eager to hop into Dr. Frankenstein's chair. The serum was infused into the test subject. The test subject not only survived, but would soon possess abilities that any military installation would kill to have. The US had its super soldier, and it was just a kid from Brooklyn.

— Steve Rogers (Captain America).

MUFON passed on Don Siedenburg's story initially, and most wouldn't touch him with a ten-foot pole. While some of his claims were certainly out there, he was able to provide me with his military records and bloodwork, which were authenticated by members of the military and medical professionals. His records were real, and his bloodwork was exceptional for someone his age.

Siedenburg's story sounds like a mix of Captain America's origin story mixed with a Roswell-like government cover-up. The fact that this story is so obscure makes you wonder why it's been dwelling in internet limbo for three years since the video interview was done.

Siedenburg claims in the interview he was given a food source from an alien spaceship as part of a military experiment. He argues the serum, which is described as thick, almost hand-sanitizer-like goo, helped him focus and increase his general cognition. He's also in great shape for an 83-year-old, as he posts basketball tutorials on his YouTube channel, and he claims he can swim a mile in around an hour's time.

He also claims to have seen crashed UFOs, dead alien bodies, which he described as mummies, and said the military used hypnosis to help hide the truth. Coincidentally, Siedenburg's military records were allegedly destroyed in a 1973 fire. He was a brand-new recruit at the time. The experiments took place at Edwards Air Force Base, California, from June 3 to 24, 1954, and is also the same place he was shown the crashed saucer and dead aliens. He went to Fort Leonard Wood Army Base, Missouri, after that for basic training.

Jordan Pease did the initial video interview (for MUFON, but they never ran with it) in 2016 and described Siedenburg as "honest and sincere, and his story deserves much more attention." The following account is from Siedenburg himself and has only been edited for AP-style grammar or clarity.

On June 2, 1954, handler Buck and I boarded a U.S. Army C-47 airplane at the U.S. Army Ordnance Depot at Savanna around noon and flew to Edwards Air Force Base in California. The distance was 1,500 miles as the "crow flies." We stopped three times before landing at

Edwards at about 5 a.m. on June 3, 1954. Buck and I were the only passengers on the plane and the flight crew numbered three. The plane was primarily used for cargo shipments and was not comfortable. When we reached Edwards the plane taxied to a hanger and we unloaded our gear.

Up to this point in our association, handler Buck had been very congenial and friendly. However, after we left the U.S. Army Ordnance Base in Savanna, I would soon find out that Buck's demeanor would change, and I was sternly given the no-nonsense details for my next three weeks during the flight to Edwards. Buck took out two pictures from his briefcase. One was the flying saucer that had crashed somewhere, and the other was two expired crew members.

The coming project that I had volunteered for was a three-week space alien flight simulation and I was the crew member. The surroundings were similar in size to the crashed spaceship. I would eat, sleep, exercise, and be a space ship crew member. I was medically monitored daily, and the hypnotist constantly reminded me of my secrecy oath. They were very concerned about claustrophobia, and other physical concerns. Handler Buck was the supervisor. I spoke to no one about this experience until I reached the age of 80.

When I was interviewing with the recruiting officer

at the U.S. Army Ordnance Base in Savanna, and I agreed to participate in the experiment, the alien food source was discussed. I assumed the food source was found in the crashed ship. Besides telling me the food source has been subjected to lab mice for a period of time, I was told the food source had never been tested on humans. They also admitted that I was the first one to qualify physically (5' 3" and 120 pounds) for the study.

The time was 8 a.m. on June 3, 1954 and the experiment began with Buck giving me tablespoon of what they called the Serum. There was no immediate reaction. The Serum had no taste, was odorless, white in color and a thick liquid. Before I was given the Serum, I had not eaten for about six hours and that was only a box lunch on the plane. They asked me if I had been hungry and what was the reaction to my first dose of space food. My hunger had almost immediately passed, and I was to remind them when or if I got hungry again.

Well, my hunger never came back during the three-week period. I was given a tablespoon of Serum every morning at 8 a.m. for the next 20 days. My thoughts about Serum was how amazing it was.

During the three weeks I was kept very busy. My exercise monitor introduced me to what they called the "Isometric Bar" found on the crashed spaceship. I worked out with this item frequently. He showed me six or seven positions to get the most circulation benefit. At the end of the three-week experiment my monitors were pleased with the results. They appeared to be surprised at how effective the Serum was.

On June 24, 1954, I said goodbye to my friend Buck, and the other monitors and I boarded a similar U.S. Army C-47 airplane for my 1400-mile trip to Fort Leonard Wood, Missouri.

Private Siedenburg spent the next eight weeks learning how to be an infantryman followed by eight weeks at the Army Administration Clerk Typist School.

After that I was sent to Orleans, France, and the U.S. Army Engineer Supply Control Center for two and a half years. I spent all my duty time as a typist and eventually became a "speed typist"! My duty station was 60 miles south of Paris.

It was surprising to me when I got 12 extra doses of the Serum after a tonsillectomy operation at an Orleans Hospital. Since I couldn't eat solid foods for twelve days after the operation the Serum fit the bill. I was also surprised to find out my friends at Edwards were still following my progress.

Siedenburg also believed that Adolf Hitler was trying to build a superweapon and reverse engineer crashed alien tech, which is a popular theory among

UFO researchers and occultists, but the "super serum" was all he could get his hands on.

There are parts of his story that are hard to believe, and there are other parts not so far-fetched.

Could this be a man's fictional depiction, or did it truly happen? You be the judge.

This book isn't advocating for "blurring the lines," nor is it promoting "slow disclosure through fiction," but rather pointing out that the lines have been blurred from the beginning of time and continue today through some of current popular culture's largest fictional institutions.

In this book, I compare some involved in the world of UFOs and study the strange as superheroes, and I do this not to make them seem larger than life, but to show that these people too are trying to change the world for the better just like mythological, historical and fictional figures have as well. I'm not saying they should be revered as Gods, or will one day be considered super-heroes; rather they are some "ordinary" people doing extraordinary things.

I tacked on two short fictional stories at the end to

show an example of how reality and science fiction always intertwine, and at times are one and the same.

We know some of these topics and cases aren't taken seriously by the mass scientific community. A wise man once told me science is great for some things, but not others. The paranormal, of which many aspects can't be studied with the scientific methods as we know them, is often ridiculed as not being plausible because it is unexplainable. It is that arrogance that laughs at topics like Skinwalker Ranch that really hurt our greater understanding of these extraordinary incidents. So if science isn't the answer and we have to depend on our own deeper understanding of consciousness, cultural associations, nonfictional texts, and allusions to history and mythos, then we still have something to try to make sense of the unexplained. The goal here isn't to discredit anomalous experiences and those who claim them, but rather to look at it from a different point of view and look at its fictional counterparts and see the common threads that exist.

If this book makes you rethink things when watching *Star Wars* or reading *The Avengers*, or makes you want to re-read Egyptian and Greek mythology, then great. The goal all along was to pique the interest and expand the imaginations of those initiated to the unexplainable, and those unfamiliar.

This being the third book under the Punk rock and UFOs moniker feels like a trilogy coming to an end. The goal of *Cryptozoology Meets Anarchy* was to show why these studies are underdog sciences, and question your own belief systems. *True Believers* touched the surface on the intersection of media, pop culture, mythos and also made an argument on why the adventure to seek truth should be taken on by all. The first two books were meant to be mass-appeal and easy to digest (the concision of the books is a testament to that), and *Stranger Than Fiction* continues that goal to reach wider audiences by showing them that these complex, "strange" occurrences aren't so strange after all.

Mike Damante is the author of the books *Punk rock and UFOs: True Believers* and *Punk Rock and UFOs: Crypto-zoology Meets Anarchy*. In 2019 he released his first YA fiction novel *Pumpkin Spice and Nothing Nice*. He continues his research at www.punkrockandUFOs.com.

Damante has appeared on radio, television and

online programs like *Coast To Coast AM* with George Knapp, *Into The Fray Radio*, Fox 26 Houston, Spectrum News Austin, *Somewhere In The Skies* with Ryan Sprague, Spaced Out Radio, Open Minds TV, Rogue Planet TV, and others.

Damante previously worked for the *Houston Chronicle* as a copy editor, writer, reporter and web producer for features, sports and news sections. He currently produces their "MIKED" music blog and has interviewed bands and musicians like Bad Religion, blink-182, Taking Back Sunday, Tom DeLonge, Tegan and Sara, Aerosmith, Donald Glover, Alkaline Trio, TSOL, and countless others. He currently teaches Journalism, English and Creative Writing, and lives with his wife and dog in Houston, Texas.